wisdom. Discover how and why it may be possible
to move freely through time and space with
consequences beyond our wildest fantasies. Realize
afresh how huge the universe of the unknown is,
and how close we may be to a leap forward that will
make the history of science to date seem like
a child's first stumbling steps.

NICK HERBERT has a doctorate in physics from
Stanford University and has written on faster-than-light
and quantum theory for such journals as
American Journal of Physics and *New Scientist.*
In addition to his consulting work, Dr. Herbert
directs physics seminars at the Esalen Institute.
His previous book was the acclaimed *Quantum Reality.*

FASTER THAN LIGHT

SUPERLUMINAL LOOPHOLES IN PHYSICS

FASTER THAN LIGHT
SUPERLUMINAL LOOPHOLES IN PHYSICS

NICK HERBERT, Ph.D.

NAL BOOKS

NEW AMERICAN LIBRARY

NEW YORK
PUBLISHED IN CANADA BY
PENGUIN BOOKS CANADA LIMITED, MARKHAM, ONTARIO

QC
24.5
.H47
1988

 NAL BOOKS TRADEMARK REG. U.S. PAT. OFF. AND FOREIGN COUNTRIES
REGISTERED TRADEMARK—MARCA REGISTRADA
HECHO EN CHICAGO, U.S.A.

SIGNET, SIGNET CLASSIC, MENTOR, ONYX, PLUME, MERIDIAN
and NAL BOOKS are published *in the United States* by NAL PENGUIN INC.,
1633 Broadway, New York, New York 10019,
in Canada by Penguin Books Canada Limited,
2801 John Street, Markham, Ontario L3R 1B4

Library of Congress Cataloging-in-Publication Data

Herbert, Nick.
  Faster than light: superluminal loopholes in physics / by Nick
Herbert
     p.   cm.
  ISBN 0-453-00604-3
  1. Physics-Popular works. 2. Quantum theory-Popular works.
3. Interstellar travel-Popular works.   I. Title.
QC24.5.H47 1988
530.1-dc19   88-12677
                                                                  CIP

Designed by Leonard Telesca

First Printing, November, 1988

1 2 3 4 5 6 7 8 9

PRINTED IN THE UNITED STATES OF AMERICA

FOR BETSY AND KHOLA

Contents

Introduction

"Ah Love! Could you and I with Fate conspire
To grasp this sorry Scheme of Things entire,
Would we not shatter it to bits—and then
Remold it nearer to the Heart's Desire?"
                    The Rubaiyat of Omar
                    Khayyam, *translated by Edward FitzGerald*

The idea that time is a fourth dimension—on a par with the familiar three dimensions of space—is a cornerstone of the "New Physics." But if time is really just another kind of space, why can't humans move back and forth in time as easily as they move around in space? Unlike our sense of the spatial dimensions, we experience time as a one-way street. Boxed inside the present moment's narrow vantage—like passengers on an express train hurtling relentlessly down the tracks of time—we look back to a memory of a fixed past and forward in anticipation to an uncertain future. Is this familiar way of experiencing the world everlastingly fixed or is it, like so many other human experiences, subject to change through technical innovation? Can the same science that opened our eyes to atoms, galaxies, and the rings of Saturn also help us see into the future or reach back and alter the past? Is there a time machine in your future?

The existence of a time machine would transform our lives beyond all present modes of description. No longer would there have to be a "road not taken"—we could simply travel back in time and make some other choice. And if that choice didn't work out, we could go back into the past and try again. In a time-traveling society, our actions would no longer be irreversible. Released from our formerly one-track lives, we could have our cake and eat it too.

A time machine would open every moment in history to scientific scrutiny, conferring on its users a kind of temporal omniscience. If it operated like H. G. Wells's classic science-fiction device, the time machine would provide more than mere knowledge of past events—it would actually transport its occupants bodily into an earlier time period, where they could experience past historical events directly, enjoying a potential omnipresence beyond the reach of those still caught in time's usual one-way flow.

Disembarking in the past without warning, uninvited guests from the future would be free to change crucial events, the effect of these changes rippling through time to modify the present drastically. The ability to willfully change the past amounts to a virtual omnipotence over human history. No mere temporal tourists, these doughty travelers in time. Omniscience, omnipresence, and omnipotence are the traditional attributes of divinity. The power to execute U-turns on time's one-way street would make time travelers nothing less than gods.

An avid science-fiction fan from the time I first learned to read, I have always had a casual interest in ways of transcending ordinary experience. However I first became seriously involved in time travel research while writing a book on the foundations of quantum theory titled *Quantum Reality: Beyond the New Physics*. Most people are aware that one way of going backward in time is to travel faster than light, but few know about the exotic faster-than-light (FTL) opportunities that seem to be open to particles that obey the laws of quantum theory.

Quantum theory is the modern science of the very small. It describes with unerring accuracy the behavior of elementary particles, atoms, and molecules, as well as atom-size black holes and the birth of the Universe itself (at its inception the Universe was considerably smaller than a proton). For more than 50 years, physicists have used quantum theory as a mathematical tool for describing the behavior of all forms of matter (protons, neutrons, and electrons, for instance) and the various fields (gravity, electromagnetism, weak and strong nuclear forces) that hold matter together. This theory provides the conceptual basis for computer chips, lasers, and nuclear power plants and has been flawlessly successful at all measurable levels.

In addition to their ordinary light-speed-limited interactions such as gravity and electromagnetism, quantum particles can influence one another in a radically new way—called the "quantum connection"—more akin to voodoo than to everyday Newtonian modes of union. The quantum connection—a blatantly superluminal link that persists between any two particles that have once interacted by ordinary means—was discovered 60 years ago, but at that time physicists dismissed this ultrafast connection as a mere theoretical artifact, existing only in the mathematical formalism, not in the real world.

In the late sixties, however, Irish physicist John Stewart Bell proved that the quantum connection is more than a theoretical construct: He showed that in order to explain certain experimental results in a reasonable way, physicists must invoke real superluminal links between quantum particles. Bell's theorem states, in effect, that after two particles interact in a conventional way, then move apart outside the range of the interaction, the particles continue to influence each other instantaneously via a real connection, which joins them together with undiminished strength no matter how far apart they may roam. Unlike all conventional forces, this perpetually lingering quantum influence is not mediated by a field but results from the fact that a quantum particle seems to "leave a part of itself" in everything it touches, a part to which it always retains instant access. The quantum connection is subtle, and, even today, more than twenty years after Bell's pioneer work, the precise sense in which two distant quantum systems can be said to be connected faster-than-light is still widely disputed.

One minor effect of Bell's theorem has been to spawn several amateur research efforts to design superluminal communicators based on the quantum connection. Many of these researchers have adopted fanciful names for their "institutes," including the Ansible Foundation, Davis, California (Wil Iley & Mark Merner); the Galois Institute for Mathematical Physics, San Francisco (Jack Sarfatti); Seattle Center for Superluminal Science (John Cramer); Notional Science Foundation, Boulder Creek, California (Nick Herbert); and the Lompico (California) Institute for Superluminal Applications ("Changing yesterday today for a better tomorrow"— Nic Harvard). The quantum connection as a possible faster-than-light signaling medium is also being investigated by university physicists all over the world.

My involvement with those attempting to exploit the quantum connection for FTL communication motivated me to look into other areas of physics where superluminal motion or backward-in-time communication seemed to be happening. This book is the outcome of that curiosity.

I am grateful to those who have patiently unearthed for me mountains of curious information concerning FTL and time-travel schemes as well as other exotic topics on the borders of science. In this regard, I would like to thank Henry Palka of Hartford, Connecticut, and Richard Grigonis of Harrison, New Jersey. Special thanks also go to Saul-Paul Sirag, John Cramer, Franco Selleri, Mark Merner, Wil Iley, and the late Philip Wright for their enthusiasm for faster-than-light devices and for their encouragement. I would also like to acknowledge the hospitality of Esalen Institute in providing a forum for the discussion of these ideas.

CHAPTER 1

Fence-Makers and Fence-Breakers: A Fascination with Limits

I am afraid I cannot convey the peculiar sensations of time travelling. They are excessively unpleasant.
— H. G. Wells, *The Time Machine*, 1895

The easiest way to go backward in time is to figure out a way to go faster than light. The theoretical barriers to faster-than-light (FTL) travel are formidable, but one can imagine some intrepid twenty-fifth-century inventor, commanding new sources of energy and imagination, obsessed with pushing his ramshackle craft closer and closer to the speed of light with the aim of breaking nature's ultimate speed limit. As his heavily mortgaged spacecraft nears light speed itself, the throb of 64 mighty Starthruster engines threatens to tear apart its flimsy hull. Just before the whole thing blows up, the ship's computer cuts in the dilithium drive. The ship gives one immense shudder, then finds itself enveloped in perfect silence, cruising faster than light. "Hey," gasps Intrepid Inventor, "Faster than light is nothin'. It's easier than bustin' the sound barrier!"

To the early test pilots, the sound barrier was no abstraction, but a real limit they could feel in their body, in their bones. Pilots that dared to fly close to the speed of sound felt their planes racked by terrible stresses, shaken by violent oscillations that often wrenched the controls from their grip. As the plane neared the speed of sound itself—dubbed "Mach 1" after German physicist Ernst Mach—these effects became more intense. In those days no wind-tunnel data existed for supersonic velocities, but on the basis of their experience at subsonic speeds most pilots believed that heavy turbulence at the speed of sound and beyond would eventually rip apart even the sturdiest aircraft. When American test pilot Chuck Yeager actually penetrated this formidable barrier, however, all turbulence ceased and he entered a region of utter calm. "I was thunderstruck," he recalls in his autobiography *Yeager*. "After all the anxiety, breaking the sound barrier turned out to be a perfectly paved speedway."

Speed myths of a similar sort abounded in the nineteenth century as well. The steam locomotive was the Victorian equivalent of the jetplane, moving people and goods at unprecedented speeds. Until this time, it was widely believed that trains could not travel faster than about 50 mph because of the immense tornadolike winds they would create along their paths. Although the effects of these winds could be mitigated by building walls along the tracks, other problems would be encountered. Experts such as British natural philosophy professor Dionysius Lardner predicted that the air would be sucked out of the cars at speeds in excess of 100 mph and all passengers would be asphyxiated. Ignorant of these outdated theories, Japanese commuters crowd without worry into "bullet trains" and the French calmly sip wine aboard TGVs (Trains à Grande Vitesse), passenger trains that can travel at speeds greater than 200 mph.

In 1900, Simon Newcomb, one of the most esteemed scientists of his day, achieved a permanent, if somewhat dubious, place in history by publishing a "proof" that heavier-than-air flight was impossible: "The demonstration that no possible combination of known substances, known forms of machinery, and known forms of force can be united in a practical machine by which men shall fly long distances through the air, seems to the writer as complete as it is possible for the demonstration of any physical fact to be." Newcomb stood his ground well into the twentieth century, dismissing the Wright brothers' flight as an inconsequential stunt.

A few years after Newcomb's bold proclamation, Albert Einstein, then an obscure clerk in the Swiss patent office, published a radically new theory of time and space which established the speed of light as the Universe's ultimate speed limit. Einstein's theory of relativity and his bold conjectures concerning the nature of space and time are now commonly accepted facts of modern physics—after 80 years of the most stringent tests, not a single experiment has ever contradicted the theory of relativity. Unlike Newcomb's ban on airplanes, which was based on his estimate of human engineering ability, Einstein's speed limit appears to be built into the very structure of space and time, as firm and permanent as the Universe itself.

Einstein's limit applies to the motion of everything—particles, waves, spacecraft, and bowling balls—not just to light rays. That we happen to call this limit the "speed of light" is simply a historical accident, based on the fact that electromagnetic radiation (which always travels at the Einstein limit) was the easiest ultrafast phenomenon to investigate in the nineteenth century. This limit could just as well be called the "speed of neutrinos," for neutrinos also go the limit, but neutrinos were discovered half a century after Einstein's invention of relativity.

One of the central passions of modern physics is the quest for a unified vision of nature. Einstein's discovery of a universal speed limit already counts as a type of unification, for it guarantees that any new particle however bizarre, any new field however complex, will be bound by the same ultimate speed limit as ordinary light rays. The fact that all of nature's entities must obey the same speed limit suggests that the search for more profound kinds of unification might not be in vain.

As a natural standard of velocity, the speed of light, c, provides a convenient measuring rod for both distance and time. Instead of expressing a star's distance in miles, we can measure it in "years," the time it would take light—the fastest signal in the world—to get there. (The light-year is a measure of space expressed in units of time.) By this light-based reckoning, Alpha Centauri, the nearest star, is 4 light-years away; our Milky Way galaxy 100,000 light-years across.

Light speed can be used to express time in units of space. In certain physics experiments, electronic pulses from particle detector A must reach a counter at the same time as pulses from another detector, B. But this temporal coincidence is possible only if the cables from A and from B are precisely the same length. To achieve synchronization, physicists insert or remove pieces of cable (called "delay lines") from one of the channels. These delay lines—down which a signal travels at essentially light speed—are labeled not by their physical lengths but by the time light takes to traverse them, leading to the amusing spectacle of dozens of "pieces of time" hanging on the laboratory wall. To these physicists, a nanosecond (one billionth of a second) is a piece of black cable about 12 inches long.

Light moves through space at the incredible speed of 186,000 mps, traveling from the Moon to Earth in little more than a second. At its closest approach, Mars is 5 minutes away by light beam, Saturn an hour away, and distant Pluto is 5 light-hours from earth. Since radio signals travel at light speed, signals from the Pioneer spacecraft near Saturn took at least an hour to reach Earth. Likewise, signals from ground control needed an hour's time to reach the spacecraft. Because of this delay, enforced by the finite speed of light, course changes had to be calculated carefully since they were based on hour-old data and commands would only be executed an hour after they were sent.

The most striking feature of light speed is its extreme rapidity as measured by terrestrial standards. Although the sound of thunder is noticeably delayed—about 1 sec/1000 feet—we seem to see the accompanying stroke of lightning immediately. Galileo was the first scientist to attempt to measure the speed of light. He stationed observers at various distances who then covered and uncovered lanterns, looking for a distance-dependent delay due to light's transit time. That Galileo discovered no such delay is not surprising. The Italian scientist did not know that light is so fast that in one second, it can circle Earth more than seven times. Measuring the speed of light evidently required greater distances than those between hills in the Tuscan countryside.

In 1676, Swedish astronomer Ole Romer noticed that the observed eclipse times of Jupiter's moons deviated from the predictions of Newtonian mechanics. These eclipses were always late when Jupiter was in conjunction with the Sun (farthest from Earth) always early when Jupiter was in opposition (nearest to Earth). Eclipse times differed as much as 15 minutes from Newton's predictions. Romer attributed these systematic deviations not to unknown mechanical effects but to the fact that during conjunction Jupiter's light has to travel a greater distance than at opposition. Close observations of Jupiter's moons allowed Romer to calculate the velocity of light and establish for the first time that light's speed is not infinite.

Frenchman Armaund Fizeau was the first to devise, in 1849, a way to measure the speed of light in a terrestrial laboratory. Fizeau shined a light beam through a rapidly rotating cogwheel's closely spaced gaps, changing the wheel's speed until light passed through one gap, traveled to a distant mirror, and returned in time to pass through the adjacent gap. Since Fizeau's day, many variations of his experiment have been carried out, determining the speed of the light with an uncertainty of only a few miles per hour.

Persuaded that the accurate determination of the speed of light was a matter of national importance, in 1879, the U. S. Congress appropriated money for this purpose and appointed Simon Newcomb, hapless naysayer of manned flight, to supervise the project, which was carried out by Albert Michelson, then a physics instructor at the U. S. Naval Academy, using a version of Fizeau's toothed wheel.

The most accurate measure of light's speed is made in a system in which everything measured is standing still. When two waves traveling in opposite directions meet under the proper conditions, they form a standing wave such as the waves one sometimes sees in coffee cups, which vibrate up and down but don't go anywhere. Measuring light's speed using standing waves makes use of the fact, valid for all waves—both moving and stationary—that the wave velocity is equal to the product of wavelength and frequency. An electromagnetic wave in a metal box (called a "resonant cavity") will form a standing wave when its wavelength is exactly equal to the length of the cavity. Measure the dimensions of the box and the frequency of the resonant wave and the product of these two stationary measurements gives the speed of light. This kind of measurement is particularly accurate for radio waves, because the frequency can be determined very precisely by actually counting individual cycles of vibration.

Since distances on Earth are all rather small compared to a light-second, the light barrier would seem to be of no consequence for terrestrial traffic. However one machine that may soon come up against the Einstein limit is the computer. As processing speeds become faster, the overall operation speed of new computers will be limited not by the processing rates of individual chips but by the light-speed-limited time it takes to transfer data between chips. To overcome this limit, science writer Richard Grigonis has called for a "sixth-generation computer," whose chips are wired together with FTL links. Because of the time-travel implications of FTL signaling, a sixth-generation "Grigonis machine" could print out answers to questions before the questions were asked, thus radically accelerating the pace of scientific research.

After computers, it is space travel that provides a major impetus to transcend the light barrier. The distance to the nearest star is 4 light-years. Most of the stars in the night sky are from 100 to 1000 light-years away. If you wanted to visit a black hole, the nearest candidate, Cygnus X-1, is 10,000 light-years distant. The diameter of the Milky Way galaxy—that immense whirlpool of stars including our own Sun—is 100,000 light-years.

The immenseness of the interstellar void dictates that even if a spaceship could travel at the Einstein limit, it would take several human lifetimes to visit most nearby stars, and the exploration of the Milky Way would involve almost-geological time scales. Because of the slowing down of moving clocks predicted by Einstein's theory of relativity, however, the people on board such fast spaceships might experience a galactic round trip as taking only a few months. The nearest-neighbor galaxy, called Andromeda or *M*31—the most distant object in the night sky still visible to the naked eye—is estimated to be 2 million light-years away. The light we see today left our next-door galaxy about the time the human species arose on Earth. Traveling at light speed, a starship would require the same time—2 million years—to complete a one-way journey to Andromeda.

The stars float in a cold black vacuum separated by distances that are immense by human reckoning. "The infinities between the stars frighten me," wrote philosopher-scientist Blaise Pascal. In his theological science-fiction series *Perelandra,* C. S. Lewis explains the vastness of interstellar distances as a divinely ordained quarantine mechanism for separating fallen worlds such as our own from Eden's, yet unsullied. In Lewis's stories, space travel is a kind of blasphemy, an impious attempt to join what God has rent asunder.

The emotion that moves most scientists to find means of contacting other worlds is curiosity, but the ever-increasing number of people on Earth may make colonization a more urgent motive for space travel. In *The Road to the Stars,* Ian Nicholson vividly dramatizes the slim prospects for solving Earth's population problem via space travel: "The total world population is growing at 2% per annum, and this implies that it will double every 35 years, reaching about 6 billion by the year 2000. . . . There are about one hundred billion stars in our Galaxy and it is thought that a considerable proportion of these may have planetary systems. . . . The expanding mass of human protoplasm, at its present growth rate, would inhabit one hundred billion planets in less than 1300 years! Furthermore, since the Galaxy is about 100,000 light-years in diameter, this colonization could be achieved only by faster-than-light travel." Nicholson's example shows that if our rate of growth does not decrease, then space travel, even at FTL speeds, will not solve our population problem. (That dilemma must be resolved here on Earth, preferably within the next few decades.)

The desire to explore space has spawned powerful chemically driven rocket engines and motivated scientists to speculate about more effective means of propulsion. There are certainly no FTL starships on anyone's drawing boards, but how far has modern technology actually progressed in the attempt to accelerate objects, large and small, up to the Einstein limit?

Very light particles, such as the electrons in a television set, or protons in high-energy accelerators, are relatively easy to accelerate to near light speed. The electrons that produce the video image in an ordinary color television picture tube, for instance, are traveling at about 30 percent of Einstein's limit when they hit the phosphor screen and light up the evening news. Electrons in Stanford's linear accelerator (SLAC) currently hold the world's speed record for artificially accelerated particles—they lag behind light speed by only a few parts per billion. In a race with light down the accelerator's 2-mile long vacuum tube, SLAC electrons come in a close second, lagging only 70 quadrillionths of a second behind the first-place photons. This brief interval of time is about a thousand times too small to be clocked by modern methods. For all practical purposes SLAC electrons travel at light speed.

Electrons and other minuscule particles can be accelerated exceedingly close to light speed by powerful electromagnetic fields. Also, just by turning on your flashlight you produce a beam of luminance—a shower of photons—that travels precisely at light speed. Television waves, radio, and radar, since they are just lower-frequency versions of visible light, also travel at velocity c, as does electromagnetic radiation of higher frequency—ultraviolet light, x-rays, and gamma radiation.

Getting photons and lightweight particles up to extreme speeds is one thing, accelerating macroscopic objects is a different matter altogether. If we call the velocity of light "1 lux," then the speeds that humans have been able to muster are most conveniently expressed in microluxes—one millionth of light speed—or "mikes." Table I shows various important velocities measured in both conventional units and in mikes. For instance, the speed of sound in air—the legendary sound barrier first penetrated by airman Yeager—is about 1 mike. Light travels almost a million times faster than sound.

The world's most powerful rocket, the Space Shuttle's main engine, has an exhaust velocity of 15 mikes. A planet's "escape velocity" is a measure of the difficulty of getting into space from the planet's surface. Only a projec-

TABLE I

TOP SPEED OF NATURAL AND FABRICATED OBJECTS

Object	Speed	
French TGV train	238 mph	0.35 mikes
Boeing 747	608 mph	0.91 mikes
Speed of sound	741 mph	1.10 mikes
Earth's rotation (equator)	1038 mph	1.60 mikes
Space Shuttle exhaust	2.7 mps	15 mikes
Fastest manned spacecraft	6.8 mps	37 mikes
Escape velocity (Earth)	6.9 mps	37 mikes
Voyager II	10 mps	54 mikes
Earth's orbital speed	18.5 mps	99 mikes
Fastest spacecraft	41.0 mps	220 mikes
Sun's cosmic motion	224 mps	1200 mikes
Escape velocity (galaxy)	226 mps	1200 mikes
Light speed	186,000 mps	1,000,000 mikes

tile fired at escape velocity or greater can escape the planet's gravitational field. The escape velocity is also a standard by which to judge the spaceworthiness of rocket engines. Roughly speaking, in order to leave the planet in one leap, a rocket should have an exhaust velocity greater than the planet's escape velocity. Escape can be achieved by slower engines but only by expending a disproportionate amount of fuel and/or by the use of multistage rockets.

Earth's escape velocity is 37 mikes, two-and-a-half times the exhaust velocity of the shuttle engine. This disparity between gravity's pull and our best rocket's chemical push means that Earth is a highly unfavorable site from which to begin space travel. Due to our home planet's relatively strong gravity, we must strain the resources of our technology just to lift relatively modest payloads off Earth's surface.

The fastest speed achieved by a manned spacecraft (37 mikes)—about the same as Earth's escape velocity—was clocked by the Apollo X astronauts as they returned from the first successful Moon landing. Voyager II, the first fabricated object to leave the Solar System, will cruise among the stars at 54 mikes. The fastest spacecraft on record is the unmanned German-American solar probe Helios B—which reached a speed of about 220 mikes. Halley's Comet attained a comparable speed during its recent close approach to the Sun. Comets and spacecraft fly fast by terrestrial standards, but light moves tens of thousands of times faster still.

Buckminster Fuller and other futurists have remarked that Earth is already a kind of spacecraft, carrying its passengers in a yearly orbit around the Sun at 99 mikes, one ten-thousandth of light speed. By measuring color shifts in the light emitted shortly after the Big Bang and now streaming throughout the entire Universe, scientists have been able to measure Earth's velocity relative to the cosmic background radiation.

When an observer moves against a light beam in a direction opposite to the beam's direction of motion, the color of the light is shifted toward the blue by an amount proportional to the observer's velocity; the light is red-shifted when the observer moves away from the light. This velocity-dependent color shift (called the Doppler effect after German physicist Christian Doppler, who measured the effect in sound waves) is responsible for the apparent change in a siren's pitch (the sonic equivalent of color) as a fire engine passes you in the street. When you move against a wave of any sort, its peaks and valleys go by faster. The whole wave then seems to be vibrating at a higher frequency. When you move in the same direction as the wave, you encounter fewer peaks and valleys—the wave's frequency in perceived to decrease. Due to Earth's cosmic motion, the relic radiation from the Universe's earliest moments is slightly blue-shifted in the direction of the constellation Leo, red-shifted in the opposite corner of the sky by an amount that translates to a velocity of 1200 mikes, within a factor of 800 of light speed.

This brief survey of speed records in the heavens and on Earth shows that all large objects, whether manufactured or natural, travel at speeds considerably slower than light. Engineers associated with the British Interplanetary Society have estimated that rockets powered by nuclear fusion could produce exhaust velocities as high as 10 percent of light speed and attain cruising speeds of the same order of magnitude. Spacecraft capable of reaching the nearest stars within a human lifetime could probably be built with modest extensions of our present technology, but no one has the slightest idea how we might amass and direct the tremendous energies required to propel ships at say, 99 percent of light speed, let alone how to actually surpass the light barrier by conventional means.

In the following chapters we examine certain areas of physics where superluminal (faster-than-light) effects make an appearance, and evaluate these effects for their ability to propel large objects past the light barrier (FTL travel), or to push small objects or disembodied "information" past the Einstein barrier (FTL signaling). The distinction between FTL travel and FTL signaling is important. If we can get large objects to go faster than light, we can certainly use such objects to carry messages. But it does not follow that superluminal message technology can be harnessed to drive FTL spaceships.

Science-fiction writers routinely invoke superluminal travel to speed up a plot otherwise slowed to snail's pace by the vastness of interstellar distances. In the popular *Star Trek* series, for instance, the starship *Enterprise*'s "warp-drive engines" propel it at speeds as high as "Warp 8," or 512 times the speed of light, the speed of a warp-drive vessel being equal to the cube of the warp factor. Warp 8 can be used only briefly, in emergencies. The *Enterprise*'s maximum cruising speed is Warp 6 (216 times light speed). At Warp 6, travel time to the nearest star (4 light-years distant) is about a week. To the crew of the *Enterprise,* warp drive technology makes star systems seem as close together as Pacific islands in the days of clipper ships.

Superluminal technology such as that portrayed on *Star Trek* would have more serious consequences than shorter interstellar travel times. Einstein's special theory of relativity shows how to turn FTL travel into travel into the past, how, in other words, to build a time machine with an FTL spaceship, or a "time telegraph" (for sending messages into the past) with an FTL communicator.

We shall see in the next chapter that Einstein's theory of relativity by itself does not prohibit superluminal signals. In fact, the theory of relativity gives us the conceptual tools for building a time machine once we learn how to send and receive FTL signals. Relativity actually shows how to make a very fast signal go backward in time. Since time machines would have paradoxical consequences of the kill-your-own-grandfather variety, most physicists believe that such machines are logically impossible; they think that Einstein's theory should be modified to exclude these time-travel possibilities. To eliminate time machines based on superluminal connections and still retain the practical successes of relativity, physicists add an extra assumption to Einstein's theory, which forbids certain kinds of FTL connections. I examine this extra assumption, and some of its loopholes, in the next chapter.

In the movie *Superman*, the Man of Steel arrives too late to save Lois Lane from death in an earthquake. Ignoring the vow made to his parents never to travel faster than light, Superman accelerates past the Einstein barrier. The film shows Superman flying around the globe faster and faster. At a certain speed, Earth stops in its tracks, and begins spinning backward, a visually effective rendition of the time-travel effect of faster-than-light velocities.

Superman and his gifted comic-book cousins represent a perennial theme in science fiction, the desire to transcend normal modes of being. One such mode is our perception of the passage of time. The human experience of time is like a river journey into the future at a particular rate—so much physical time takes up so much psychological time. One second per heartbeat is a crude measure of how fast time seems to flow for us. Plants, animals, or extraterrestrials might feel time passing at a slower or a faster rate, as flowing backward, or even as multiple streams, but lacking a telepathy machine that opens up the inner life of other beings, time for us seems destined to flow inexorably in a single direction, out of a fixed past, into an uncertain future.

We can, of course, retrieve the past in the form of memories or via various technical analogues of human memory: films, books, and tapes. In a certain sense, television's "instant replay" is a kind of time travel: We can peer into the past and watch the winning touchdown a second or a third time.

An impressive example of memory-based time travel is holographic interferometry. First a holographic image is made of an object at noon. The holographic plate looks like a window through which a three-dimensional image is seen floating in space. When the plate is moved, the image moves with it. The plate is now moved until this noon image exactly overlaps the object itself. In regions where the object has slightly changed its shape in the interval between exposing (noon) and superposing (now), a pattern of light and dark zebra stripes covers the object, a result of light-wave interference between the object as it was at noon, and the object as it is right now. In this process an object in the present is made to interfere with a three-dimensional image of itself as it was in the past. Holographic interferometry conjoins past and present events in a particularly strange and intimate way. Although holographic interferometry and instant replay do in some sense bring the past into the present, these techniques do not achieve true time travel because they only retrieve records of past events; they do not bring back or allow us to change the past events themselves.

Because the flow of time is basically a psychological experience, many science-fiction stories invoke psychological mechanisms for time travel rather than mechanical time machines. Such mental vehicles have included hypnotism, consicousness-altering drugs, mental illness, and, in a class by itself, masturbation (Ray Nelson's "Time Travel for Pedestrians"). But the bulk of science-fictional time travel, following the lead of H. G. Wells's pioneering *The Time Machine,* utilizes some sort of physical mechanism for transcending time's otherwise inevitable flow.

In the following chapters we'll consider some physical mechanisms that might be employed for time travel (or for time telegraphy) not because they manipulate time's flow directly, but because they involve superluminal connections which Einstein's relativity theory predicts would give rise to circumstances in which the past itself could be reexperienced, not merely its records. Before discussing the superluminal possibilities of specific areas of physics, we briefly survey Einstein's relativity theory, looking at why physicists believe that FTL travel is impossible; how relativity actually supports the notion of time travel; and how FTL travel could be converted to time travel if the Einstein limit can be circumvented.

CHAPTER 2

Is Time Travel Possible? The COP Says No; Minkowski Says Maybe

The mattermitters of Sector Yellow and the Burst, or the leisurely lightpushers that are popular in remote regions like Sector Gray or Violet, carry between them only fifteen percent of the galaxy's traffic. Small ferry ships and freighters . . . account for another eighteen percent. The rest of the tonnage, goods or passengers, plunges through phase space in FTL (faster-than-light) ships.

—Brian Aldiss, *Starswarm*, 1964

For American physicist Albert Abraham Michelson, measuring the speed of light was more than a job, it was an obsession. From his days as a midshipman in the U. S. Navy to his retirement from the faculty of the University of Chicago, Michelson devoted his entire life to devising ingenious methods for clocking nature's fastest phenomenon, light itself. In 1880 he invented the Michelson Interferometer, an instrument for measuring tiny distances using a pair of light waves as a kind of Vernier caliper, a technique that more than a century later is still unsurpassed in accuracy. As a tribute to his remarkable accomplishments in the field of optics, a recent biographer dubbed Michelson "The master of light." Working in a field dominated by Europeans until the middle of this century, Michelson became the first American to win a Nobel prize in Physics.

In the 1880's, when Michelson began his science career, the biggest problem in physics was the nature of the ether, a hypothetical medium in which light supposedly moved.

Just as sound waves need air to travel in, and ocean waves need water, so physicists assumed that light waves must travel in a medium called the luminiferous (light-carrying) ether, which is present everywhere, even in so-called "empty" space.

In still air, sound travels about 700 mph. However, if you drive your boat toward a foghorn at 50 mph, the horn's sound will seem to move past you at 750 mph. Likewise if a 50 mph wind blows between the foghorn and your stationary boat, the sound waves, aided by the wind, will move past at 750 mph.

If light waves in the ether behave like sound waves in the air, then accurate measurements of the speed of light in various directions should be able to detect the movement of this ubiquitous light-carrying medium past the observer. Light-speed measurements could, in principle, reveal the presence of an "ether wind." As part of his goal to determine light's velocity accurately, Michelson set himself the task of measuring the magnitude and direction of Earth's motion through the ether.

How fast might we expect the ether wind to be blowing past Earth? In Chapter 1, we express the velocities of various objects in microluxes (1 microlux = one millionth of light speed), or mikes. (Calling this light-based velocity unit the mike might also be construed as honoring Albert Michelson for his lifelong devotion to the task of measuring the speed of light.) The major motions of Earth consist of its orbit around the Sun, and its cosmic motion as a member of the Milky Way galaxy. As previously mentioned, Earth's orbital velocity is about 100 mikes, and its cosmic velocity is about 1200 mikes.

From this information we might reasonably expect the speed of light to vary by as much as 1200 mikes, slowed down if the light is traveling against Earth's motion through the ether, speeded up if the ether wind happens to act as a "tail wind." Since the direction of Earth's orbital motion changes during the year, we might also anticipate a seasonal variation in the ether wind to the extent of 100 mikes or so.

To measure the ether wind, Michelson, together with chemistry professor Edmund Morley, set up in Cleveland, Ohio, a kind of optical racetrack (now called the Michelson Interferometer), that pitted light bouncing back and forth between two mirrors in a north-south direction against a similar light beam traveling in the east-west direction. Depending on the direction of the ether wind, one of the light beams has the track advantage and is sure to win.

To this day, the Michelson-Morley method represents the most sensitive way for searching for an ether wind. Morley and Michelson's optical racetrack was sensitive enough to detect an ether wind as slow as 25 mikes—four times slower than Earth's orbital velocity. Despite this high sensitivity, the Michelson-Morley experiment and all subsequent experiments always resulted in a photofinish between the two light beams. No matter in what direction the interferometer is aimed, light always seems to travel at the same speed. Also there are no seasonal variations in the speed of light when measured at different times of the year.

Some physicists tried to explain the utter absence of an ether breeze by imagining that the ether gets caught up and carried along with the motion of Earth, forming a kind of ether atmosphere fixed to Earth's surface. According to this "ether drag" hypothesis, because the ether sticks to matter, the true ether wind could only be measured far from Earth. Attempts by Oliver Lodge and others to observe the ether drag near massive rotating wheels in the laboratory were unsuccessful. Furthermore, the abrupt change in the ether's velocity between Earth and outer space should distort starlight as it travels to Earth's surface, altering the apparent position of the stars. But no such shifts in stellar position have ever been observed.

Faced with this new data, physicists devised increasingly ingenious schemes for reconciling the notion of an ether with the negative results of the ether wind experiments. Dutch physicist Hendrik Antoon Lorentz and Irish physicist George Francis FitzGerald independently proposed that when something moves through the ether, the electrical forces holding it together are modified in such a way that an object shrinks in the direction of motion; matter moving through the ether experiences a kind of "ether squeeze." Since all measuring rods also contract by the same amount, the "Lorentz-FitzGerald contraction" is never directly measurable. In ether wind experiments such as those of Michelson and Morley, the apparatus shrinks in the direction of the ether wind just enough to compensate for light's slower speed in that direction fooling the experimenter into believing that light travels the same speed in all directions. Lorentz and FitzGerald tailored their contraction to have precisely the right magnitude to cancel out the effect of the ether wind. Physicists rightfully regarded this mysterious coincidence—two unmeasurable quantities that just happened to cancel out exactly—with suspicion. Yet such theoretical contortions seemed necessary if one wanted to retain both the idea of an ether and the experimental fact that light speed is measured to be the same in all directions.

At the height of the ether-modeling frenzy instigated by Albert Michelson's experimental results, the light-speed controversy was resolved in 1905 by another Albert—Albert Einstein—in a strikingly novel manner. Einstein simply abolished the ether and replaced it with the concept that space, time, and mass measurements are relative to an observer's motion. According to Einstein, there is no absolute time and space, the same for all observers. Instead my space and time is different from your space and time, if we are moving relative to one another. As Einstein put it in his first paper on relativity: "The introduction of a luminiferous ether will prove to be superfluous inasmuch as the view here to be developed will not require an absolutely stationary space." Insulated from the physics mainstream, Einstein carried out research in the hours he could spare from his duties at the Zurich patent office. No doubt this professional isolation was partly responsible for the impressive originality of his thinking.

Einstein recognized that the search for the ether was equivalent to a search for an absolute frame of reference, for a way to determine, in empty space, who is really moving and who is standing still. Physicists regarded the ether as the physical embodiment of Isaac Newton's notion of absolute space, that universal stillness against which Newton and his followers calculated the motions of the stars and the planets. "Absolute space," declared Newton, "in its own nature, without regard to anything external, remains always similar and immovable."

According to Newton, one special space is at rest; all others are really moving. In the quiet of the Zurich patent office, Einstein wondered what sort of world would result if the notion of absolute space were discarded so that every observer could with equal right assert that he was the one who was standing still. What it means for all frames to be equivalent, Einstein decided, is that the laws of physics must be the same for all observers. Einstein revolutionized physics by tossing out all privileged reference systems—including the cherished ether.

Not all motion is relative in the Einsteinian picture, only motion that is uniform in magnitude and direction. Accelerated motion is not relative, because the laws of physics are different in accelerated systems, making it possible to tell which frame is accelerated and which is not. Rotation, for instance, is a type of accelerated motion in which the direction of motion is constantly changing. The rotation of Earth leads to peculiar forces, called centrifugal and Coriolis forces.

Centrifugal force is an outward-directed force resulting from rotation—the force that tends to pull you off the merry-go-round. At the equator, where Earth is moving the fastest, people actually weigh a few ounces less than at the poles because of centrifugal forces generated by Earth's rotation. The Coriolis force caused by the rotating Earth acts on all moving bodies, deflecting them to the right in the northern hemisphere, and to the left in the southern hemisphere. Some scientists contend that it is the Coriolis force that makes drain water in a bathtub spin clockwise above the equator, counterclockwise below. Others maintain that the Coriolis force is too small, compared with other forces in the bathroom environment, to enforce a particular rotation direction on your dirty bathwater. Because of the occurrence of these extra forces—which are never present in a stationary frame—we can say with certainty that Earth is rotating, not the starry sky. No such acceleration-dependent forces distinguish two rockets moving past one another with uniform velocity. Einstein's first relativity postulate states that there is no meaning to the question "Who is really moving?" in such a case. The laws of physics are exactly the same in both rocket ships.

As the first building block of his new theory, Einstein assumed that the laws of physics are the same for all uniformly moving observers. But what is meant exactly by "the laws of physics"? In Einstein's day, physics was founded on two sets of laws, Scottish physicist James Clerk Maxwell's laws of electromagnetics and Isaac Newton's laws of motion.

An elementary consequence of Maxwell's laws is that the speed of light is a constant, c, whose numerical value is fixed by the ratio of electric to magnetic forces. In 1861 Maxwell actually calculated the speed of light from the measured force between two standard magnets and the electric force between two charges. These forces do not change when an observer moves past the charges and magnets, so the speed of light, as calculated from Maxwell's theory, always has the same numerical value—186,000 mps—no matter what the observer's state of motion might be. Newton's laws, on the other hand, predict that a light beam should travel at different speeds for observers who themselves are traveling at different speeds. Newtonian physics asserts that light travels at speed c only for an observer who is absolutely at rest.

For the numerical value of the speed of light, these two laws—the Mosaic tablets of classical physics—gave two different results; they could not both be right. Einstein believed more in Maxwell's laws than Newton's and claimed as his second relativity postulate that the speed of light is constant for all observers. Einstein's choice meant however that Newton's laws are wrong. Einstein therefore replaced Newton's laws with new laws of motion—called relativistic mechanics—according to which the speed of light is constant for all observers. Since Newton's laws embody many common sense ideas concerning motion, Einstein's new mechanics had revolutionary consequences for our ordinary notions of space and time.

Einstein's relativity principle states that the laws of physics—that is, the way individual measurements relate to one another—must be the same for everyone, but this does not mean that everyone's measurements of the same event will be the same. Most of my measurements of an event—if I happen to be moving relative to you—will look different from yours, but when I fit all my measurements together into a certain pattern, both our patterns will agree. These invariant patterns are the laws of physics, the same for all observers. Physics consists largely of the search for such patterns—ways of putting together observations that disagree to form structures that are the same for everyone.

Suppose I am moving at one-half the speed of light toward a distant star, and you are standing on Earth. According to Einstein, the speed of the starlight is the same for both of us. The speed of light is one of these invariant patterns. As part of Maxwell's laws of electromagnetics, which Einstein's relativity left unchanged, the constancy of light's speed has the status of a law of nature. Now the only way that light speed—the measured distance that light travels divided by the measured time it takes to go that distance—can be the same for both of us if our notions of time and space are different. The gist of relativity is this: In order to keep the speed of light constant, my perceptions of space and time must be different from yours. Einstein's theory is called "relativity" because according to his theory space and time are not absolutes, as Newton and common-sense supposed, but relative to an observer's state of motion.

From his two postulates Einstein derived a set of relations that describe how space, time, and mass seem to change in a moving reference frame. These relations—the heart of special relativity—are called the Lorentz transformations, after Dutch physicist Hendrik Lorentz who derived them prior to Einstein in connection with a certain mechanical model of the ether. Although Einstein did not use the concept of the ether in his reasoning, he came up with the same relations Lorentz did for the different perceptions of space and time experienced by observers moving at different velocities.

The Lorentz transformations predict four major changes that objects in a moving frame seem to undergo as viewed from a fixed frame: (1) Space shrinks in the direction of motion (Lorentz contraction); (2) time slows down (time dilation); (3) clocks desynchronize (relativity of simultaneity or "sync shift"); and (4) mass increases in the moving frame.

To visualize these four consequences of Einsteinian relativity, imagine that you and your friend Max are in a rocket ship traveling at 90 percent of light speed with respect to another ship piloted by Max's twin sister Maxine. According to the first postulate of relativity, any observer has the right to regard himself as standing still, and to regard everybody else as moving. Since you and Max, from your point of view, are standing still, you do not expect to see any change in your length, your time, or your mass due to your rocket ship's motion. According to you the one who's moving is Maxine.

The first rule of thumb of relativistic mechanics is that nothing strange ever happens to the guy who's standing still; it's always the other guy—the one who's moving—whose physical reality seems cockeyed. Of course the same relativity rule allows Maxine to regard herself as the one who's standing still, and to say that all the weird stuff is happening to you and Max. Surprisingly, the Lorentz transformations, which relate Max's viewpoint to Maxine's, allow both twins to hold different opinions concerning the same space-time events without contradiction.

When the relative velocity of the two ships is small compared to the speed of light, both Max and Maxine agree concerning their measurements of space, time, and mass. As their relative velocity approaches light speed, however, their perceptions begin to differ by an amount proportional to the "Einstein factor." (The Einstein factor is equal to 1 for low velocities, is equal to 2 for 90 percent of light speed, increases to 7 at 99 percent of light speed, and becomes infinite at light speed itself.)

From Max's point of view, Maxine is traveling at 90 percent of light speed, a velocity for which the Einstein factor is 2. Because of her motion, Maxine's space (all her distance measurements), time (all her clocks), and mass (the response to gravity of her rocket and its contents) are modified by the Einstein factor, that is, for the case of relative motion at 90 percent of light speed, all Maxine's measurements will either be doubled or halved compared to Max's. In particular, the length of Maxine's rocket ship seems contracted by a factor of 2: Maxine's ship appears to Max to be only half as long as it was in port. Likewise when Maxine looks back at Max with her laser radar, she considers herself to be standing still, and accordingly sees Max's moving ship diminished in length by a factor of 2.

Although this relativistic length contracted (sometimes called the Lorentz contraction—leaving out poor FitzGerald) is numerically equal to the old Lorentz-FitzGerald (L-F) contraction, its physical interpretation is entirely different. The L-F contraction was supposed to have arisen from an actual physical interaction between matter and a stationary ether. According to Einstein's theory, on the other hand, this contraction is not due to any physical property of matter but results from a heretofore unsuspected property of space-time itself. From a fixed frame, the structure of a space that is moving will seem to have contracted in the direction of motion. Any matter that happens to occupy this space will likewise seem contracted.

In concert with length contraction, time contracts as well. Max sees all of Maxine's clocks running slow by the same Einstein factor that affects her lengths. From Max's viewpoint, while 10 sec tick off in his rocket, only 5 sec seem to go by in Maxine's ship. Not only Maxine's physical timepieces, but all cycles mechanical and biological are slowed by the same factor. Like the Lorentz contraction, this slowing of clock rates—called "time dilation"—is symmetric: From Maxine's point of view, it is Max's cycles that are running slow.

In addition to altered perceptions of one another's lengths and clock rates, Max and Maxine each observe a third relativistic effect—relativity of simultaneity—an effect that is crucially important for all time-travel schemes involving faster-than-light motion. Unlike the previous effects, which do not depend on the direction of the spaceship's motion, the sync shift effect for approaching ships is opposite in sign from the sync shift for receding spacecraft. If Max and Maxine are approaching each other at 90 percent of light speed and Maxine's ship (by Max's reckoning) is 1.1 light-years away, then Max will swear that Maxine's clocks are all set ahead by 2 years. When Max's calendars say 1990 Maxine's say 1992. This difference in clock synchronization occurs in addition to any synchronization differences that accumulate due to different clock rates.

Unlike the time dilation effect, which depends solely on relative velocity, the sync shift effect increases with distance. At a separation of 2.2 light-years, Maxine's clocks are 4 years ahead of Max's; at 4.4 light-years, her clocks are 8 years ahead. Doubling the distance between ships doubles the sync shift.

The time dilation effect (slowing of clock rates) is independent of direction. Maxine's clocks run slow whether she is receding or approaching. However the sync shift has a different sign, depending on the direction of relative motion. In particular when Max and Maxine are separated by

1.1 light-years and receding from one another, Maxine's clocks are all 2 years behind Max's. When Max's calendar says 1990, Maxine's say 1988. The sync shift operates to make the moving clock look like it's set ahead when the twins are approaching one another, and appear to be set behind time when the twins are receding. At the moment when the twins whiz past one another, neither receding nor approaching, their clocks are perfectly synchronized.

The sync shift effect—which because of relative motion Maxine's clocks are not synchronized with Max's clocks—means that it is impossible to define consistently a "present moment" valid all over the Universe. If you are moving, your definition of "now" will be different from mine just as Max's "now" is different from Maxine's. Thus when Einstein eliminated Newton's absolute space, he abolished Newton's absolute time as well. That time is no more with us of which Newton declared, "of itself and from its own nature, flows equably without regard to anything external, and by another name is called duration." In Einstein's universe, every relatively moving observer possesses his or her own time each with its own set point (What is now?) and its own clock rate (How fast is time passing?).

The sync shift is particularly important for superluminal signaling schemes because, as we shall see later in this chapter, it is two observers' differing accounts of what constitutes "the present moment" that permits certain FTL signals to travel backward in time.

There is a sense in which the relativistic length contraction is "not real" but is merely a consequence of the sync shift effect. In order to measure a moving object's length you must note the position of its front and back ends "at the same time." Imagine, for instance, trying to measure the length of a rapidly swimming goldfish. Observers who fundamentally disagree about what is meant by "the same time" will necessarily mark the tail and head positions differently and hence calculate a different length for the same goldfish.

Max and Maxine each perceive the other's space and

time to be distorted by the effects of relative motion. In addition, each sees a fourth relativity effect—the mass of the moving ship and its contents have increased by the Einstein factor. To Max's mass probes, Maxine's 1000-ton ship now seems to weigh 2000 tons. Maxine, of course, denies any such weight gain, claiming that Max is the one whose mass has doubled.

This relativistic mass increase constitutes the first barrier to FTL travel. If you try to accelerate a body up to the speed of light, it gains mass as it approaches the Einstein limit, and hence presents increased resistance to further acceleration. At the speed of light itself, the body would have infinite mass—immovable no matter how large the applied force. Near light speed, most of the energy put into a body goes into increasing its mass, very little into increasing its velocity. Because of the relativistic mass increase it is impossible to accelerate a particle to a velocity equal to that of light, let alone to go faster. Superluminal speeds cannot be achieved by brute force; more subtle means are needed.

Inside a color television picture tube, an electronic gun about 4 inches long accelerates electrons to 30 percent of light speed, about 300,000 mikes. For these fast particles, the Einstein factor is about 1.05, which means that the electrons that light up a television screen are 5 percent heavier than their slower cousins elsewhere in the Universe. Physicists Edwin Taylor and John Wheeler have drawn attention to an important economic consequence of the Lorentz transformations. The Stanford linear accelerator is designed to speed electrons to near light speeds. If an electron's mass did not increase with velocity, the accelerator would need to be only slightly longer than a picture tube and would cost about the same as a good television set. However, passage through the SLAC increases each electron's mass by an Einstein factor of 50,000, a bullet traveling as fast as a SLAC electron would weigh as much as a dump truck. To accelerate particles this massive one needs a lot of energy. To get the SLAC electrons moving at their top speed (99.999999992 percent of light velocity) the accelerator has to be 2 miles long and would cost $300,000,000. Thus the size of the SLAC budget is a direct consequence of Einstein's theory of relativity.

In a rare display of technological heraldry, the insignia of the American Physical Society depicts the three fundamental units on which all of physics is based: time (symbolized by a pendulum); length (symbolized by a ruler); and mass (symbolized by a brass weight). Ironically, relativity has taught us that none of these standards of physical measurement remains constant when viewed from a moving reference frame. What does remain constant, the same for all observers, are certain fundamental constants such as the speed of light and the laws of physics, which include relativistic mechanics, electromagnetics, and quantum theory. A scientific heraldry more in tune with relativity would emblazon the symbols of these invariant laws on its shield rather than the inconstant quantities that are represented at present.

Although the relativistic mass increase prevents brute force attempts to push matter past the light barrier, it does not hinder more artful strategies. As we shall see in Chapter 7, quantum theory offers more subtle possibilities for putting a particle on the other side of the light barrier without actually passing through the region of infinite mass.

Quantum theory contains many options that would be unthinkable in a classical world. For instance, a quantum particle such as the positron can be created out of sheer energy, or an electron can tunnel from one allowed region to another allowed region through an intervening region where classical physics says no particle should be able to go. Thus under the new quantum rules, the mass barrier might be circumvented by creating a particle that travels FTL at the moment of its birth, or by arranging somehow for a particle already traveling very close to the Einstein barrier—such as a SLAC electron—to simply quantum-tunnel to the other side. (We discuss in Chapter 7 the possibility of creating FTL particles called tachyons by utilizing these quantum loopholes.)

When confronted with the almost magical powers of quantum theory, the relativistic mass increase barrier to FTL travel might seem rather leaky. Also this barrier does not outlaw "teleportation," in which an object simply disappears at one location and instantly reappears somewhere else. No velocities, superluminal or otherwise, are involved in teleportation, yet the object achieves effective FTL travel. To plug these leaks and clearly establish all notions of superluminal transport as heretical, physicists added an extra assumption to Einsteinian relativity, an assumption we call the causal ordering postulate, or COP.

Special relativity does not in itself outlaw superluminal motion. What relativity does say is that certain kinds of superluminal motion lead directly to time travel, that is to signals that can go back into the past, signals that are capable of changing events that, by conventional reckoning, have already happened. To eliminate the possibility of time travel via superluminal signaling from the laws of nature, physicists attempted to make the weakest assumption possible that would do the job—the COP. This extra postulate cannot simply outlaw all superluminal motions because, as we shall see in Chapter 3, certain processes occur in nature that are known to go faster than light. The job of this new postulate is not to eliminate all FTL processes, just those that can be used to build time machines.

According to common sense and the laws of Newtonian mechanics, no matter how fast a signal travels, it never goes backward in time. In Einsteinian relativity, however, FTL signals sent between two events in space-time will appear to some observers to travel in a reversed temporal order. To understand how the COP works, consider the three different ways that two events can be separated in Einstein's world. When time and space are both measured in light-speed units (for instance, time = years, space = light-years), three kinds of separations are possible for two events, A and B, in space-time. If B can be reached from A

by a slower-than-light signal, the separation is called "timelike" because for such slow signals more time than space stretches between A and B. If B can be reached from A by a ray of light, the A-B separation is called "lightlike." Lightlike separations span equal amount of time and space. Finally, if B can be reached from A only by FTL travel, the separation is called "spacelike." More space than time is spanned by such superluminal separations.

Timelike separations can be bridged by vehicles or signals that travel at less than light speed. Two events that are lightlike-separated can be connected only by a signal that travels at the speed of light. Both timelike and lightlike separations can be linked up by conventional signals. A spacelike separation, on the other hand, is too far apart to be connected by anything slower than a faster-than-light signal. In physics parlance, the term "spacelike-separated" means a pair of space-time events whose linkup necessitates an FTL signal; to a physicist the word "spacelike" is synonymous with "faster-than-light." Any pair of events in space-time belong to one of these catagories: timelike, spacelike, or lightlike.

If the Sun suddenly turned green (space-time event A) at 12 noon, the inhabitants of Earth would first see the green light (space-time event B) at 12:08, because the Sun is 8 light-minutes distant from Earth. Because events A and B are linked by a light beam (green in this case), these two events are lightlike-separated. Suppose event A is again the greening of the Sun, but event B is some time on Earth later than 12:08. The events A and B are timelike-separated because information concerning the Sun's greening can be conveyed to event B by slower-than-light means.

Now let event B be any time on Earth between 12 noon and 12:08. Then events A and B will be spacelike-separated because it would require an FTL signal for someone on Earth to become aware of the Sun's noon greening at any time before 12:08. One important case of spacelike separation concerns events that happen at different places at the same time. For example, event A—the Sun's noon greening—

and event *B*—noon on Earth. To connect such simultaneous events requires an instantaneous connection. The instantaneous connection (infinite signal velocity) is the fastest FTL signal imaginable. (When discussing FTL communicators I will assume for the sake of simplicity that they can produce instantaneous signals.)

Now suppose that event *A* bears a causal relation to event *B*: *A* is the cause; *B* is the effect. For instance, *A* might be the throwing of a switch; *B* the lighting of a lamp connected to the switch. Or *A* might be the sending of a radio message; *B* the reception of that message. Or more to the point, *A* might be Max stepping into a teleportation booth; *B* is Max emerging from a distant teleportation chamber. The COP is the assumption that the temporal order of cause and effect is the same for all observers. Thus if a cause *A* precedes its effect *B* in Max's reference frame, then *A* must precede *B* in Maxine's frame too, no matter how fast or in what direction she is moving.

Because of relativity, space and time undergo strange distortions when viewed from moving frames. A pair of events (tick-tock) in one frame may seem to happen in reverse order (tock-tick) in another frame. The COP assumption does not outlaw all events that possess ambiguous time ordering. COP merely says that if two events are causally connected, their time ordering must be unambiguous. Out of all the pairs of events occurring everywhere in the Universe, COP singles out only those pairs connected by causal links, and declares that the temporal direction of this causal influence must appear the same to all observers. Thus if Max on Earth throws a switch that opens his garage door on Mars, he perceives that the Earth-switch event happens first, the Mars-door event happens later. For this situation the COP assumption requires that all observers agree which event was earlier in time, although they may disagree over how many seconds went by before the door opened.

A curious aspect of COP is that it does not require that a cause always precede its effect. COP actually permits backward causality, in which an event in the future influences what happens in the past. However if such backward-causal events actually occur in nature, COP requires that they appear backward-causal in all reference frames.

The practical consequences of special relativity—that is, the apparent distortion of space, time, and mass due to relative motion—are all contained in the Lorentz transformations, which Einstein derived from only two assumptions: (1) There are no privileged reference frames and (2) the laws of physics are the same for all observers. Strictly speaking, COP is not part of relativity but is an extra assumption concerning the presumed nature of causal relations in space-time. This extra assumption provides, unlike the relativistic mass increase, a virtually leakproof barrier to FTL travel and/or FTL signaling.

To understand how COP outlaws FTL processes, we look at how the Lorentz transformations act on the three basic types of space-time separation. Applied to timelike or lightlike separations, the Lorentz transformations never change the time-ordering of event pairs. For these kinds of separations, if A precedes B in one frame, A will precede B in all frames. However for spacelike separations the situation is quite different. A Lorentz transformation can always be found that will reverse the time-ordering of any pair of spacelike separated events. This means that if A precedes B in one frame, there will always be another frame in which B precedes A. As a consequence of the rules of special relativity, the time-ordering of spacelike-separated events is always ambiguous. Consequently the COP assumption leads to the conclusion that spacelike-separated events can never be causally connected. No signal can connect such events; no traveler can traverse them as part of his journey. Thus the COP assumption in conjunction with the Lorentz transformations permits noncausal FTL connections but outlaws all faster-than-light connections that can actually be used to change things, all situations in which a cause can produce an effect across a spacelike separation.

*　　*　　*

Faced with the formidable COP barrier, a determined time traveler has only two alternatives: (1) He can simply ignore the COP assumption and accept the consequence that in certain situations causal ordering simply gets ambiguous, or (2) he can devise FTL connections that are effective but not strictly "causal."

Most of the FTL travel plans discussed here, including tachyons (hypothetical superluminal quantum particles) and space warps, tacitly assume that the COP assumption simply does not apply in certain unusual situations, however the quantum connection discussed in Chapter 8 is more subtle. The quantum connection may be an example of an effective FTL link that does not violate COP. A causal connection is by its very nature asymmetric: There is an arrow pointing from cause to effect. COP requires this arrow to always point in the same direction. Imagine a new type of connection in which two events mutually influence each other's behavior in a symmetric way, without either event being singled out as cause or effect. Superluminal connections of this symmetric sort—which I call "sympathetic" rather than "causal"—are not forbidden by COP, and may indeed be the sort of FTL links that nature routinely deploys in the quantum world.

The COP assumption was made to explicitly outlaw FTL signaling. But why is FTL signaling so dangerous? Isn't it true that no matter how fast you go, you can never go backward in time? Now we will see how to make a time machine having the ability to signal into the past by using two FTL radio stations in different reference frames.

It is indeed true that no matter how fast Max sends his signals, none of them can ever go backward in time—for Max. However, because of the relativistic sync shift effect, under certain circumstances Max can send a signal backward in time in Maxine's frame. Such a backward signaling opportunity arises when Max and Maxine are receding from one another at high speeds. Suppose these space traveling twins are moving away from one another at 90 percent of light speed, and are presently 1.1 light-years apart. Max in the year 1990 sees, because of the sync shift, that Maxine's calenders say 1988—2 full light years behind his own way of telling time. If he sends an instantaneous signal it always stays in 1990 in his frame but it arrives at Maxine's frame in 1988. Because of the same sync shift effect, Maxine, as she receives Max's signal in the year 1988, sees that Max's calendars say 1986: from Maxine's point of view, it is Max who is 2 years behind. If Maxine now returns Max's signal instantaneously, the signal will arrive back at Max's ship in 1986. The overall effect of this instantaneous signal exchange between two receding frames is that Max sends a signal in 1990 and receives it back in 1986. Using Maxine as a relay, Max is able to signal 4 years back into his own past!

In this example, Max and Maxine used instantaneous signals, but slower signals will work—as long as they are sufficiently faster than light. The faster the rocket velocities, the slower the signals can be and still go backward in time. No matter how fast the frames are receding, however, a slower-than-light signal exchange will never go backward in time.

With only one FTL radio it is impossible to signal backward in time, but two such radios, mounted in mutually receding reference frames, would constitute an actual time machine, a way of influencing the past from the vantage point of the present. However, the combination of special relativity (Lorentz transformations) and COP outlaws FTL signaling and seems to make time machines of this type impossible as long as one abides by the relativity/COP rules. Ironically, despite the formidable conceptual barrier to time travel, another aspect of special relativity strongly suggests that time travel is not an unreasonable possibility. The major motivation for time travel in fact stems from a way of looking at relativity devised by Hermann Minkowski, one of Einstein's physics teachers at the Zurich Polytechnical Institute.

Minkowski noticed that the Lorentz transformations formally resemble a mathematical rotation, a rotation not however in ordinary space but in a four-dimensional space which, in addition to the ordinary three spatial dimensions, incorporates time as an extra dimension. In Minkowski's view, the length contraction and time dilation are analogous to the illusion of foreshortening that an ordinary object presents when rotated in ordinary three-dimensional. Many people had speculated before Minkowski that time is the fourth dimension, for instance the popular writer C. H. Hinton (*What Is the Fourth Dimension?*, 1880) and novelist H. G. Wells (*The Time Machine,* 1895) but Minkowski was the first person to justify the concept of fourth-dimensional time by showing how it is implied by the laws of physics.

Minkowski's insight shed light on why physical laws are invariant for moving observers, but space and time are not invariant. Physical laws are the same for all because they are expressed in terms of four-dimensional mathematical objects called "tensors." Just as ordinary objects retain their true form under rotations, although their dimensions seem to shrink and expand according to the rules of visual perspective, so laws constructed from tensors retain their form when subjected to uniform motion, the four-dimensional equivalent of three-dimensional rotation.

Space and time by themselves are not tensors since they appear different for different observers. However a certain way of combining these quantities such that the relativistic changes in time exactly cancel the changes in space gives rise to a quantity called "invariant interval," a tensorial quantity for which all observers report the same value. The "invariant interval" is the four-dimensional equivalent of the "real length" of a rotating stick that appears foreshortened by various amounts depending on its angle with respect to an observer.

Minkowski's version of relativity as a four-dimensional physics fixes the shape of all future scientific laws. If it's not formulated in tensor form, it will not look the same to all observers, and consequently cannot possibly be a correct law of nature. Today no physicist considers his theory complete until he has been able to express it in tensorial form, as a set of relations between four-dimensional objects. Because of his clever formulation of Einsteinian relativity as a four-dimensional physics—permitting an otherwise abstract theory to be visualized in simple geometric terms—Minkowski has been described as "the man who understood Einstein."

In his famous address to the 80th Assembly of German Natural Scientists and Physicians in Cologne in 1908, Minkowski introduced his new interpretation of relativity this way: "The views of space and time I wish to lay before you have sprung from the soil of experimental physics and therein lies their strength. They are radical. Henceforth space by itself, and time by itself, are doomed to fade away into mere shadows, and only a kind of union of the two will preserve an independent reality." Minkowski's intimate new union of space and time—the arena in which the New Physics operates—is termed space-time.

Although Einstein initially regarded Minkowski's innovation unfavorably, he soon learned to think in four-dimensional terms. Taking the notion of space-time seriously led Einstein to the greatest discovery of his scientific career, the general theory of relativity. Starting with Minkowski's flat space-time, Einstein introduced the more general notion of space-time curvature. In 1915, he announced a radically new theory of gravity in which gravity is no longer a force but a measure of the local space-time curvature.

General relativity now forms the basis for our understanding of all gravitational phenomena including the motions of the planets, the expansion of the Universe, and the birth of black holes. The beauty and success of Einstein's theory add considerable weight to Minkowski's claim that despite its three-dimensional appearances, the world we live in is in reality four-dimensional.

The most radical consequence of Minkowski's four-dimensional viewpoint is that the past still exists (and so does the future). Minkowskian space-time is a kind of frozen snapshot of eternity. The predictive success associated with this peculiar space-time viewpoint is a strong motivation for believing in the possibility of time travel. If the past in some sense still exists, then one can seriously consider schemes for paying it a visit.

For the physics of time travel, Einstein's relativity theory provides both the motivation—the permanent presence of the past in Minkowskian space-time—and the method—FTL connections plus relativistic sync shifts. Using relativity as a road map, we have learned that, to take a trip into the past, all you have to do is find or make two systems, A and B, that are receding at a rapid but subluminal speed from one another, then devise a way to send an FTL signal from A to B, and to return an FTL signal from B to A., (Pay no attention to that COP who says such signals are against the law.) The return signal (B to A) will return before the original signal (A to B) was sent, the backward temporal displacement increasing with the relative recessional speed of systems A and B. This ability to signal backward in time depends on the ability to achieve two-way signaling at faster-than-light velocities. In the following chapters we'll investigate apparent superluminal phenomena in various fields of physics and the suitability of these phenomena as components of a time machine.

CHAPTER 3

Things That Go Faster Than Light

"What is an ansible, Shevek?"
"An idea." He smiled without much humor. "It will be a device that will permit communication without any time interval between two points in space. The device will not transmit messages of course; simultaneity is identity. But to our perceptions that simultaneity will function as a transmission. . . ."
—Ursula Le Guin, *The Dispossessed*, 1974

The first barrier to brute force FTL travel is the relativistic mass increase; subtler forms of FTL travel—such as direct teleportation or quantum jumps—are blocked by the COP rule, which outlaws all spacelike causal connections. In spite of these relativistic prohibitions, however, there are still things in the world that travel faster than light. Indeed relativity itself guarantees FTL travel of a certain sort.

The fact that a body's mass increases with its speed was discovered experimentally by Wilhelm Kaufmann, in Berlin, a few years before Einstein published his relativity theory explaining this mass increase as an effect of moving observers' different perceptions of space and time. As a particle's velocity increases, its mass increases by the Einstein factor, necessitating a proportionately larger force to achieve the same acceleration. As the particle's velocity approaches the speed of light, the Einstein factor approaches infinity. Infinite mass means that no force, however large, can make the particle go any faster. Kaufmann first observed this velocity-dependent mass increase for electrons, which are the lightest particles known, excepting those with zero mass. Zero-mass particles (called "luxons") always travel at light speed and can neither be accelerated nor slowed down. At present only three luxons are known: the proton, the neutrino, and the hypothetical graviton.

Because of its tiny mass, the electron is the easiest particle to accelerate. In high-energy accelerators, electrons travel only slightly slower than light itself. At Stanford's linear accelerator the electron's speed differs from light speed by less than 1 mph. Some physicists have suggested, somewhat tongue in cheek, that this small velocity gap could be closed by moving the entire accelerator. The speed of the accelerator added to the electron's near-light velocity would combine to produce electrons traveling at light speed or faster. For instance, a circular accelerator might be mounted on a turntable, making a kind of physicist's merry-go-round by which one might hope to sneak under the light barrier by adding from outside a small velocity to a particle already traveling at nearly light speed.

If, for example, an electron is traveling around a ring-shaped accelerator at 90 percent of light speed we might think about rotating the ring itself at 20 percent of light speed, hoping to get the electrons to travel at 110 percent of light speed. However near-light velocities don't add up in the usual way. For this kind of situation the rules of relativity would produce an electron velocity only 93 percent of light speed.

The relativity rules give the same disappointingly slow result no matter what scheme we use for adding velocities. If a rocket ship traveling at 20 percent of light speed ejects an escape pod at 90 percent of light speed, the resultant pod velocity will be only 93 percent, not 110% of light speed, just as in the rotated accelerator case. In the extreme case where the rocket is traveling at almost the speed of light, and shoots off a light beam, the resultant light does not travel at twice light speed but merely at light speed. Here relativity leads to the unusual conclusion that one plus one equals one.

* * *

For high-speed space travel, a more mundane barrier than the relativistic mass increase plagues the would-be astronaut—an effective limit on a rocket's speed is set by its own exhaust velocity. All rockets operate by expelling a substance backward. This backward loss of momentum then causes the rocket to acquire an equal momentum in the opposite direction. The speed of ejection of this backward moving substance is called the "exhaust velocity." A ship can attain a velocity higher than its own exhaust only by a prodigious expenditure of fuel. When a rocket's speed exceeds its exhaust velocity, the process of acceleration becomes increasingly inefficient and fuel consumption rises exponentially. Unless there is a willingness to expend a disproportionate amount of fuel, a spaceship captain cannot get the ship to go much faster than its own exhaust. The liquid hydrogen/liquid oxygen fuel used in NASA rockets can achieve an exhaust velocity of about 15 mikes—light speed is 70,000 times faster. No chemically propelled rocket can go much faster than about 50 mikes and such a rocket would take 80,000 years to reach the nearest star.

Rockets propelled by thermonuclear reactions—controlled hydrogen bombs—rather than chemical fuel have been proposed, the Daedalus project of the British Interplanetary Society is one example. Daedalus is a spaceship designed for a trip to Barnard's star—6 light-years away—which exhibits a wobble in its motion indicative of the possible presence of planets. A thermonuclear reaction is capable of producing an exhaust velocity of about 10 percent of light speed (100,000 mikes); the design speed of Daedalus is 16 percent of light speed (160,000 mikes), which would mean a one-way trip of 40 years to reach Barnard's star.

The most efficient relativistic rocket would have an exhaust velocity equal to the speed of light. Its exhaust would consist of particles accelerated to near light speed or of light itself or some other luxon. Present methods of turning fuel into light, such as lasers, or accelerating particles to relativistic speeds are quite inefficient. The only known 100 percent efficient process for turning matter into light is matter-antimatter annihilation, but no one has any idea how to use the annihilation process in a spacedrive.

In 1960 Robert Bussard of the TRW Corporation in California proposed using the interstellar medium itself—which contains a few hydrogen atoms per cubic centimeter—as a rocket fuel. The fact that the ship's fuel is obtained from the outside eliminates the exhaust-velocity limit; the Bussard jet is capable of efficiently accelerating to speeds greater than its own exhaust. The Bussard ramjet would employ an enormous scoop constructed out of electromagnetic fields to collect and funnel hydrogen into a nuclear-reaction motor. The Bussard jet actually becomes more efficient at high speeds because the faster it goes the more hydrogen it collects. Many science-fiction writers have used Bussard ramjets in their flights of fancy, however the Bussard principle suffers from the fact that at high ship speeds the fuel is not standing still but impacts the ship's scoop at enormous velocity. The incoming fuel constitutes a huge headwind, which becomes increasingly difficult to overcome as the ship goes faster. The speed of the Bussard jet soon reaches an upper limit—about 10 percent of light speed—where thrust is equal to wind resistance and the ship can accelerate no further.

Another barrier to relativistic travel is the enormous energy needed to accelerate even modest payloads to near-light velocity. To accelerate a spaceship to 90 percent of light speed (where its space, time, and mass are modified by an Einstein factor of 2) requires an amount of energy at least equal to the spaceship's rest mass. To get a Volkswagen bug to go this fast would require total conversion of almost a ton of matter into energy—equivalent to the amount of all the energy consumed by the United States in a year.

Although energy and efficiency problems seem difficult to overcome, they can—and are—open to technological breakthroughs. Perhaps an unlimited source of energy will be discovered. Putting these merely practical problems behind us, we would then confront the speed-of-light barrier itself, which no amount of energy can surmount. Surprisingly, relativity itself describes a way in which a ship could in a certain sense travel faster than light without ever crossing the light barrier.

By a curious numerical coincidence, the product of Earth's gravitational acceleration, g, and the number of seconds in a year is approximately equal to the speed of light. This means that a spaceship that could produce a steady acceleration of 1 g would take only 1 year to reach relativistic velocities. In addition to getting up to light speed in a reasonable amount of time, this 1-g spaceship would provide its passengers the familiar feel of Earth-like gravity.

Let's go for an imaginary trip in a 1-g spaceship. Table II charts our ship's progress over a 25-year (ship time) voyage. After 1 year, the ship's velocity as seen from Earth (call this the "real velocity") is 0.71c. Through the rear windows the ship's passengers see Earth receding at the same velocity. Because of ~~Einsteinian~~ *LORENTZIAN* time dilation, folks on earth will perceive the ship's clocks to be running slow (reduced by the ~~Einstein~~ *KAUFMANN* factor). Using this slow clock, and Earth units of length to calculate velocity, your ship's captain reckons that he is traveling precisely at the speed of light. Although this way of calculating speed does not play fair because it divides Earth length units by rocket-ship time units, it does represent the measure of velocity most relevant to the space traveler. If a star is 5 light-years away (as measured from Earth), and you can get there in exactly 5 years (5 of your subjective years, that is), then as far as you are concerned, you have covered the distance at light speed. Using this "practical" measure of velocity, it is not only possible to achieve light speed, it is easy to travel faster than light.

The stay-at-homes on Earth and the ship's passengers each have an explanation for the ability of the 1-g ship, apparently, to exceed the speed of light. They explain this effect in different ways, but relativity works in such a way that both explanations are correct. The folks on Earth perceive the distances between the stars as normal, but say that because the spaceship is moving, the ship's clocks and all other onboard periodicities (heartbeats, metabolism, meal times, and menstrual cycles) are slowed down by the Einstein factor. It is the fault of these slow clocks, say the folks on Earth, that the ship's passengers believe they can travel faster than light. By the same kind of reasoning, you could claim to run a 26-mile marathon in 1 minute, if you timed your run with a very slow clock.

On the other hand, the folks on the ship consider themselves standing still and claim that their clocks (and all their biological cycles) are running at their normal rates. These passengers, however, see the stars rushing past at a terrific velocity. Because of their high speed, the Lorentz contraction shrinks the distance between stars by the Einstein factor. If you are traveling at near light speed and the distance between stars has been reduced by a factor of 10, then you will seem to be traversing space at an effective speed ten times greater than the speed of light.

The stay-at-homes explain the ship's apparent superluminal velocity as a consequence of time dilation inside the ship. The voyagers explain their high effective speed as a result of the Lorentz contraction of the space outside. Both explanations are correct and both explanations give the same (effectively superluminal) velocity.

Table II shows that when 5 spaceship years have gone by, almost 75 years have passed on Earth. Because all cycles on board ship are slowed by the Einstein factor, the space travelers actually age at a slower rate than Earth dwellers. When the space-faring folks return, relativity predicts that they will be many years younger than their stay-at-home cousins. Recently, John Hafele and Richard Keating, at the U. S. Bureau of Standards in Washington, D. C., actually reproduced this differential aging effect by flying identical atomic clocks in opposite directions around the world on commercial airlines. (A big part of their experimental budget consisted of airline tickets.) Because of Earth's rotation, the west-to-east clock was traveling almost 1000 mph

TABLE II

FLIGHT PLAN OF A 1-G ROCKET

Ship Time	Earth Time Years	"Real" Speed Years	Practical Speed
0.7	1.0	$0.71\,c$	$1.0\,c$
1	1.18	$0.76\,c$	$1.18\,c$
2	3.62	$0.96\,c$	$3.62\,c$
5	74.2	$0.9999\,c$	$74.2c$
10	11,000	$0.99999+\,c$	$11,000\,c$
20	500 million	$0.99999+\,c$	500 million c
25	72 billion	$0,99999+\,c$	72 billion c

faster than the east-to-west clock. When the clocks were reunited, the faster clock was discovered to be slightly slow— having aged less than its identical partner by a fraction of a microsecond. Because of the slow speeds involved, transcontinental air travel is no shortcut to longevity. You would have to take 25 million around-the-world trips (eastward) to lose a single second of time relative to people on the ground.

At relativistic speeds, the age difference between Earthlings and astronauts can amount to several centuries: your family and generations of their descendants would be long dead before you returned. In her science fiction stories, Ursula Le Guin connects distant worlds with NAFAL (nearly-as-fast-as-light) ships that travel at "real speeds" slightly less than light but at "practical velocities" many times the speed of light. Le Guin's heroes jump quickly from star to star at the price of forsaking forever family and friends. Leaving on a NAFAL ship is a departure as final as death; one is condemned to wander forever among strangers because your home world will have aged by centuries while you remain young. As a consequence of this relativistic fountain of youth, NAFAL travelers tend to be gloomy types, aloof and disinclined to make close friends.

*   *   *

Checking back into our 1-g spaceship we see that after 20 shiptime years our ship has attained a practical speed 500 million times the speed of light. At this velocity, it would be possible to cross the entire Milky Way in less than 2 hours, about the time it takes to fly from Chicago to New York. Just 5 years later the ship is going more than 70 billion times light speed, and could cross the entire Universe in a matter of a few months. Only 25 years have passed in the ship but the Universe itself has been aging at its normal rate and is now 72 billion years old.

In some models of the cosmos, gravity eventually pulls the Universe back together into a final "Big Crunch," a time-reversed mirror image of the initial Big Bang. The present age of the Universe is about 20 billion years by most reckonings, perhaps a third of the way through its cosmic cycle. So, the passengers on the 1-g ship might actually be able to witness the Big Crunch courtesy of special relativity, a possibility dramatized by Poul Anderson in *Tau Zero* (Tau is the universe of the Einstein factor—when the Einstein factor gets big, Tau goes to zero). In Anderson's story, a Bussard ramjet provides the enormous speeds required to carry his characters off to the end of the world.

From the example of a 1-g starship we see that relativity itself provides a way to exceed, "in practice," the speed of light when velocities are calculated with distances and times taken from two different reference frames. But are there things in the world that travel FTL when velocities are calculated in the proper manner?

Consider a pair of scissors. As they close, the notch where the two blades meet moves out toward the scissor's tip with a velocity that increases as the angle between the blades gets smaller. When the blades become parallel, this notch velocity is infinite. For very long scissors, the velocity of the intersection can exceed that of light during the last instant of closure. Because the scissor's blades are massive, they must always move slower than light, but the intersection of the blades is massless, a mere geometric concept, and is not subject to any relativistic mass increase. Can the Einstein speed limit actually be exceeded by common cutlery?

Imagine a pair of scissor blades extending to the moon (1-1/2 light-seconds away). If we close the blades in 1/2 sec, we send a signal moonward at three times the speed of light. One is reminded of Archimedes, who boasted that given a long enough lever (and a place to stand) he could move Earth. Given two levers, in the form of giant scissor blades, a modern Archimedes might well brag about his ability to break the light barrier.

Alas, this scissor-blade signaling scheme depends on the hidden assumption that the blades are made of a material that is infinitely rigid, so that when I make a move on Earth, the entire blade moves at once. Such infinitely rigid objects do not exist in our world. If such materials were available, it would be simpler to replace the scissors with a single perfectly rigid rod. Push this rod's Earth end and its Moon end moves immediately. An infinitely rigid rod all by itself could be used as an FTL communicator. In reality, all rods are flexible, and deformations in such rods move at the speed of sound. For the purpose of this experiment, it is as if the scissor blades were made of rubber—a sudden movement on Earth takes a long time before it is felt on the Moon.

However, despite the flexibility of the blades, it is still possible to get the intersection moving at superluminal speeds. If we station along each blade little rocket ships timed to fire at the proper instant, we can get each blade to move and the scissors to close, as if it were a rigid body. The intersection will then move faster than light, but we cannot use this motion for sending signals because this motion does not originate on Earth—it is made possible only because of preplanned events not just on Earth but everywhere between Earth and the Moon. Although the intersection moves FTL, this motion is not so much a message as a formal "demonstration" carried out by well-rehearsed players stationed between Earth and the Moon.

* * *

Another similar FTL "demonstration" is a series of lights that blink in sequence like the "moving" lights on the rim of a theater marquee. By proper programming, these lights can appear to move at any velocity, even velocities greater than light. But it is easy to see that nothing really moves on the marquee, and certainly no message can be sent along this row of blinking lights. Synchronized scissors and marquee lights are examples of FTL motion not prohibited by the COP assumption. The COP assumption does not outlaw superluminal motions or connections as such. What COP does forbid are FTL motions and connections that can act as a medium for causal influences.

A light display in which light #1 triggers light #2 which triggers light #3 would be an example of a causal connection. COP says that this type of light show cannot ever move faster than light. However if we drive all lights from a single source programmed to turn them on at selected times, the lights are no longer causally linked to one another, COP does not apply and any light pattern is possible, including patterns that ripple along at superluminal speeds.

Another example of COP-permitted FTL movement is a searchlight beam. As the beam sweeps across the sky, the end of the beam (like a massless version of Archimedes's lever) moves faster and faster as it gets further away. At a certain distance, the beam is actually traveling faster than light, but if we look at what is actually moving we see that the end of the light-beam resembles the marquee lights—in both cases a light flashes somewhere, and later another light flashes somewhere else. The human brain connects the two independent flashes and produces an illusion that "something" jumped from one place to another. Psychologists have studied this illusion of motion and shown that if the second light flash is delayed too long—more than a second for most people—the brain will not produce the "jumping light" illusion but will correctly perceive the phe-

[handwritten margin note: THE LIGHT BEAM WOULD MOMENTARILY BE "BENT"]

nomenon as two independent lights. This FTL "searchlight effect" can take many forms. When a comet swerves close to the Sun, its tail, which may extend for millions of miles, streams out like a searchlight beam pointing directly away from the Sun. As the head of the comet rounds the Sun, the end of its far-flung tail can, in principle, exceed the speed of light.

For the same reason that a searchlight beam can travel FTL, so also is it possible for a shadow—a sort of negative searchlight—to exceed light speed. For example, the shadow of a very fast planet, close to its star but far from a second planet, can race across the second planet's surface with a faster-than-light velocity. Hence some alien life forms dwelling on a planet of this kind may occasionally witness the spectacle of a superluminal eclipse.

Another marquee-like example of FTL motion is the electron beam in an oscilloscope or television picture tube. Like the searchlight beam, the writing speed of an oscilloscope —the speed at which the luminous dot moves across the screen—can, in principle, exceed the speed of light. In practice, the fastest commercial scope, the Techtronix 7104, achieves a writing speed only 60 percent of light speed. By simply doubling the length of the 7104's display tube, Techtronix engineers could build a superluminal oscilloscope whose spot velocity would be 120 percent of light speed. Some experimental scopes probably already exceed the light barrier but since no violation of the COP is involved, scope engineers don't make a fuss about "superluminal oscilloscope traces." However the manufacturers of the next generation of oscilloscopes might want to celebrate their new instruments by claiming "our scopes can write faster than light."

Similar to the FTL motion of scissors and oscilloscope traces are the "galloping waves" discovered by a group of physics students at Georgia Tech. When two waves collide head on, the resulting wave forms can be quite complex, and sometimes include transient wavelets that for brief moments actually move faster than light. These Georgia Tech galloping waves were not actual waves in some physical system but were produced and clocked on a computer.

Galloping waves can be compared to "riptide," another potentially superluminal phenomenon connected with wave motion. When an ocean wave strikes the beach at an angle, the breakpoint where the wave meets the beach moves sideways along the beachfront at a velocity that depends on the wave's angle of approach. When the wave hits the beach at right angles, the velocity of this "riptide" is infinite—the wave breaks simultaneously all along the beach. Clearly the riptide phenomenon is similar to scissor blades: No signal can be carried along the beach by this ultrafast ripple. The superluminality of galloping waves stems from the same sourcé as that of riptide, except in this case two waves are colliding rather than one wave and a beach. Like superluminal marquee lights, galloping waves arise from a sort of coordinated conspiracy and are useless for FTL signaling.

Apparent FTL situations can also result from mistaking certain absolute motions for relative motions. For example, as viewed from Earth, the heavens seem to rotate once every 24 hours. This means that in order to circumnavigate Earth in such a short time, some of the very distant planets and stars must go exceedingly fast. To argue that, since all motion is relative, Earth can be considered to be at rest, then the stars must be moving at superluminal speeds, is an incorrect application of the first relativity principle, which states that every observer is entitled to regard himself as being at rest. This principle applies only to observers moving with velocities uniform in both magnitude and direction with respect to one another. For systems moving with accelerated motion, such as our 1-g spaceship, or the rotating Earth (whose constant speed continually changes its direction), it is possible to say for sure who is accelerating, and who is not. Acceleration is an example of absolute motion, not a relative motion. It is the 1-g rocket that is accelerating not Earth; consequently one twin ages, the other does not. It is Earth that is really spinning, not the stars. Consequently, despite their apparent diurnal rotation, the stars do not actually go faster than light.

In ancient cosmologies, Earth was believed to be stationary, circled daily by the moving stars. From these cosmologies the notion of astrology was born, granting psychological attributes and influences to the celestial bodies based on their appearance and behavior in the sky. Now that the distances to these celestial bodies are known, it is possible to calculate each planet's apparent velocity due to its daily motion across the sky. The result of this exercise in geocentric thinking is that nearby planets always travel more slowly than light. The two most distant planets, however, Neptune and Pluto, always move at FTL velocities, in keeping perhaps with their association in astrology with the occult, the imagination, and fantastic inventions.

Moving outside the orbit of Pluto, far beyond our Galaxy and its near neighbors, we enter the realm of the mysterious quasars, whose real nature is still controversial. Quasars are very distant, intensely luminous objects that appear to emit the power output of an entire galaxy from a region no bigger than our Solar System. Using an array of radiotelescopes, astronomers can resolve the structure of nearby quasars and have recorded their changing shapes. They look like tiny glowing amoebas exuding writhing pseudopods into space. Remarkably, the outlying parts of many quasars (at least eight to date) seem to be expanding at superluminal velocities ranging between two and ten times the speed of light. Astrophysicists currently explain this apparent FTL expansion as an optical illusion that depends on special geometric alignments and the finite velocity of light. If one component, *A*, of the quasar is considered stationary, and the other, *B*, is moving toward us with relativistic speed at a small angle to the line of sight, then the light from the *B* component has a headstart compared to light from component *A*. For the same reason you can see a baseball being hit and only later hear it, the real velocity relations between components *A* and *B* are distorted by the finite travel time of light.

To produce the illusion of FTL expansion, however, the quasar orientation must be quite special. If quasars are oriented at random, conditions for apparent superluminality should occur in only 1 to 2 percent of all cases. Thus if we find an increase in the number of superluminal quasars, the optical illusion explanation will soon seem quite contrived.

Astronomers may be able to use these "superluminal" quasars as a rough measure of the size of the Universe because their apparent FTL motion is directly linked to our present astronomical distance scales. The apparent speed of a quasar's expansion depends on how far away we estimate these objects to be: The further away the quasar is assumed to be, the faster component *B* will seem to be moving. If on closer examination of other quasar parameters, the FTL expansion is too fast to be explained as an optical illusion, perhaps our present cosmic distance scales need to be revised to reduce these FTL expansions to reasonable magnitudes. So far, all quasars analyzed in this way have been explainable as optical illusions, suggesting that our present distance scales are not wildly mistaken.

Quasars are the most distant objects that we know about in the Universe. They are visible on Earth only because of their exceptional brightness. The quasar light that reaches us today was emitted billions of years ago. This immensely ancient light gives us a direct look at phenomena that were taking place when the Universe was very young.

Although light travels at the same speed no matter what the motion of its source, such motion can shift the light's frequency. Light from a receding object seems red-shifted (lowered in frequency); approaching objects are blue-shifted (raised in frequency). One of the most distinctive features of quasar light is its large red shift due to the expansion of the Universe. Our present model of the Universe begins with a Big Bang when the Universe was expanding at its maximum rate, then a gradually decreasing rate slowed by gravitational attraction. Einstein's general theory of relativity describes this cosmic expansion not as a giant explosion of matter but as an expansion of space itself, which carries all matter along with it. Whether the mass of the Universe is large enough to halt the expansion and to cause the cosmos to contract eventually is an unanswered question. Present estimates of the Universe's mass turn up an amount of material a factor of 30 short of the critical mass needed to turn the Universe around. If this present estimate is correct, then the Universe is destined to go on expanding forever, at a gradually decreasing rate.

Because space was expanding faster in the past, light rays emitted from distant objects are stretched (red-shifted) by an amount proportional to their travel time (very old light is stretched the most). A quasar's red shift is a measure of the age of its light, how long the light has been traveling, hence how far away the quasar was when the light began its long journey to Earth.

We see the oldest quasars at a time when the Universe's expansion rate was its greatest. The greatest of these expansion rates are in fact equivalent to velocities several times greater than the speed of light. If the Universe were expanding at a constant rate we could never hope to see such distant quasars because their light would be in effect "swimming upstream" against a current that was moving faster than the light itself. We can see these distant objects only because as time passed the expansion rate slowed down, allowing the light to make headway against the current.

Special relativity requires that all objects embedded in space obey the light-speed limit, but it places no restriction whatsoever on the speed of expansion (or contraction) of space itself. Despite the fact that large portions of faraway space are routinely expanding faster than light, there seems to be no way to exploit this universal stretching of space-time to send superluminal signals from one part of space to another except in science fiction. Kurt Vonnegut's ingenious Tralfamadoreans, in *The Sirens of Titan*, achieve FTL space travel by hitching their starships to "the Universal Will to Become," an allusion perhaps to titanic processes such as spatial expansion which affect the entire cosmos.

The superluminal stretching of space due to the overall expansion of the Universe took place at a time far in the distant past. News of this ancient FTL velocity comes to us only indirectly via the abnormally large red shift of quasar light. Closer to home and more directly observable is the behavior of radio waves in the upper atmosphere. Around the turn of the century, pioneer radio engineers discovered that during part of their trip from transmitter to receiver these waves actually move at FTL velocities. When a beam of light passes from one transparent medium to another it does not usually travel in a straight line but bends at the interface to an extent and in a direction that depends on the ratio of light velocity in the two media. The direction in which light bends depends on whether it is going from a high-speed to a low-speed medium or vice versa. When light travels from a space where its speed is fast into a medium where its speed is slower, the beam tilts down deeper into the second medium. ("Fast into Slow; dives down below.")

On the other hand, when a light-beam travels from a medium (such as glass) where its speed is slow, into a medium (such as air) where its speed is fast, the beam tilts up, away from the straight path, as it penetrates the second medium. ("Slow into Fast; takes the upward path.") For an incident angles greater than a certain "critical angle," the light-beam for a slow-into-fast situation, instead of bending up, doesn't even penetrate the second medium, but reflects from the interface as if from a perfect mirror. This phenomenon of "total reflection" is characteristic only of the slow-into-fast situation and never occurs in the fast-into-slow case.

Since water is a slow medium (75 percent of vacuum speed) compared to air (99.9 percent of vacuum speed), light that travels from water into air is liable to be totally reflected at the interface at shallow angles of incidence, that is, for certain angles the air-water surface behaves like a perfect mirror. The phenomenon of total reflection can be observed by divers and dishwashers. In the first case, when viewed from below, the surface of the water turns into a brilliantly shimmering reflector except for a "hole in the mirror" directly overhead through which the above-water world is visible. In the second case, the inside of an air-filled glass immersed in dishwater turns into a perfect reflector—it looks like the glass is full of mercury—because light trying to pass from a slow medium (glass) into a fast medium (air) is totally reflected.

Radio engineers were aware of the Einstein speed limit so they were astonished at the discovery of a medium that totally reflected radio waves, because they knew that these waves were already traveling at the speed of light. Since total reflection occurs only when waves travel from a slow medium into a fast medium, the occurrence of such reflection could only mean that radio waves in the second medium must travel faster than light.

This superluminal medium for radio waves is nothing more exotic than an ionized gas—a gas whose atoms have been broken up into positive and negatively charged particles called "ions." Neon signs and fluorescent lamps are filled with ionized gas, also called "plasma." The plasma state is often referred to as the "fourth state of matter" comparable with the more familiar solid, liquid, and gaseous states. Although examples of plasmas are sparse on Earth, the plasma state is by no means rare. Since the Sun and stars are made of ionized gas, the plasma state is the most common form of matter in the Universe.

How can the fact that radio waves in a plasma travel faster than light be reconciled with the Einstein speed limit? Throw a stone into the center of a quiet pond. The stone's impact creates a circular disturbance that moves out from the center at a certain speed, a disturbance that takes the form of a gradually expanding ring. If you look closely at this expanding band, you will see that the band consists of little ripples that are moving faster than the band itself. These ripples start at the back of the band, move forward at about twice the band's speed, then disappear at the band's leading edge. Thus there are actually two velocities connected with a water wave, the velocity of the ripples (called "phase velocity") and the velocity of the band itself (called "group velocity"). It is obvious in this case that the speed of the ripples has nothing to do with how fast the water wave gets across the pond.

Whenever a radio pulse travels through a plasma, the velocity of the pulse (group velocity) always obeys the Einstein speed limit, but the velocity of the ripples on the radio pulse always travel faster than light. Most scientists believe that radio signals can only be sent at the slow group velocity, that it is impossible, as in the water wave case, to use the FTL ripples to send a message. If the phase velocity does not carry a signal, it cannot act as a link in a causal chain and hence is exempt from the COP prohibition. In Chapter 2 we learned that by itself Einstein's theory of relativity does not outlaw FTL motion. However when the causal ordering postulate is added to Einstein's theory, all FTL processes that can carry a causal influence—signals or bulk transport through space, for instance—are forbidden. Since most of the superluminal processes discussed in this chapter are obviously unsuitable as carriers of causality—they are simply illusions of motion, variations on the moving marquee-light theme—such processes are COP-allowed, but the same lack of causality in these phenomena renders them useless as components of an FTL-based time machine. Most of these superluminal processes are useless for communication purposes but the question of whether phase velocities can be harnessed to carry a signal is more subtle. In the next chapter we consider the case of the superluminal phase velocity in more detail.

TABLE III

FOURTEEN THINGS THAT MOVE FASTER THAN LIGHT

Scissor-blade intersection	Marquee lights
Searchlight beam	Comet tail
Eclipse shadow	Riptide
Perfectly rigid rod	Oscilloscope trace
Galloping waves	Neptune and Pluto
Quasar expansion	Expansion of space-time
Plasma phase velocity	"Practical speed" of NAFAL ship

CHAPTER 4

Phase Waves: Superluminal Vibes in the Upper Atmosphere

"In a nutshell," Weinbaum said, "ultrawave is radiation, and all radiation in free space is limited to the speed of light. The way we hype up ultrawave is to use an old application of wave-guide theory, whereby the real transmission of energy is at light speed, but a quasi-imaginary thing called phase velocity is going faster. . . ."
—James Blish, *"The Quincunx of Time,"* 1973

In the winter of 1901, camped out on Labrador's frozen shore, young Italian inventor Guglielmo Marconi detected a weak radio signal—the letter "S" in Morse code. This little message, traveling 2000 miles across the Atlantic from a station in Cornwall, England, shattered forever 5 billion years of radio silence on Earth. Today radio waves crisscross the globe, pour down from satellites, and reach out via radio telescopes to the very edge of the Universe. We move every day in an immense ocean of electromagnetic signals which protect, guide, inform, and entertain us, an electromagnetic babel whose first utterance was Marconi's little letter "S." As the radio age dawned, Marconi's name became a household word. Less well-known is the puzzle Marconi's long-distance radio reception posed for the physicists of his day, a puzzle that bears on Einstein's speed-of-light limit.

The physicist's problem was to explain how Marconi's message could travel so far. Between Cornwall and Labrador Earth's curvature imposes an impenetrable barrier effectively 100 miles high. Because of their wave nature, radio signals are expected to bend to a certain extent—the familiar diffraction effect that allows sound waves to be heard around corners. Long wavelengths bend the most; short waves much less. The waves put out by Marconi's transmitter were much too short for his signal to bend via diffraction around the curvature of Earth. Marconi's waves had to have traveled in straight lines like beams of light. For the same reason that a searchlight cannot shine across the Atlantic, Marconi's short-wave signals should not have been able to reach Labrador.

The first part of the radio puzzle was solved by British physicist Oliver Heaviside, who proposed the existence of a layer of ionized gas in the upper atmosphere which could act as a mirror to reflect the waves back to Earth. Ordinary air does not conduct electricity because its molecules are electrically neutral. Air in this upper atmosphere, on the other hand, is ionized (broken into positively and negatively charged ions) by the solar wind, an erratic outpouring of charged particles ejected from the surface of the Sun. American physicist Arthur Kennelly independently proposed the same idea. In Marconi's day these gaseous radio mirrors in the sky were called the "Kennelly-Heaviside layers." Today they are simply known as the "*D, E,* and *F* layers." The atmospheric region where these clouds of ionized gas occur (100–150 miles high) is called the ionosphere.

Like the weather at lower altitudes, the height and location of these reflective layers at the interface between Earth and outer space is not entirely predictable. A favorite sport of radio amateurs (called "DXing") is to attempt to exchange messages over the greatest possible distances using the least amount of transmitter power by exploiting fortunate conjugations of these high-sky radio mirrors.

The presence of the Kennelly-Heaviside layers—later verified by rocket-borne ionization sensors—solved the mystery of long-distance, short-wave radio propagation. But a paradox still remained when physicists tried to compute the speed of radio waves in these layers of electrified gas. As discussed in Chapter 3, whenever a wave travels from a medium where its speed is slow into a medium where its speed is fast, the beam bends up, back toward the interface. At a certain critical angle of incidence, the slow-into-fast wave is bent parallel to the interface and does not penetrate the second medium at all. For all incident angles greater than this critical angle, the beam is totally reflected. For these angles the interface acts like a perfect mirror. The sparkle of a diamond is partly due to light trapped inside a slow medium by total reflection: In a diamond, light travels at only 40 percent of its vacuum velocity.

Total reflection at the interface between two transparent media occurs only when light tries to go from a slow medium into a faster medium. When the situation is reversed (fast into slow) only a small fraction of the incident light is reflected; most simply goes right through. The fact that the ionosphere totally reflects radio waves implies that radio waves travel faster in the ionosphere than in air. However, radio waves in air are already traveling at the E̶i̶n̶s̶t̶e̶i̶n̶ *Maxwell* limit. Thus the total reflection of radio waves off these invisible alphabetical layers in the sky is only possible if waves in the ionosphere travel faster than light!

The apparently superluminal velocity of radio waves in the ionosphere was also confirmed by theoretical physicists. Using Maxwell's equations, which govern all electromagnetic phenomena, it is possible to calculate the velocity of electromagnetic waves in an ionized gas. The results of such a calculation are shown in Figure 4–1 as a function of radio frequency.

Below a certain frequency—called the "plasma frequency" ("plasma" is another name for ionized gas)—the gas completely absorbs all waves. For low-frequency waves, the plasma looks black. Just above the plasma frequency, the wave velocity is infinite, but rapidly decreases with frequency, approaching the speed of light at very high frequencies. At all frequencies for which wave propagation is possible, the speed of radio waves in a plasma is always

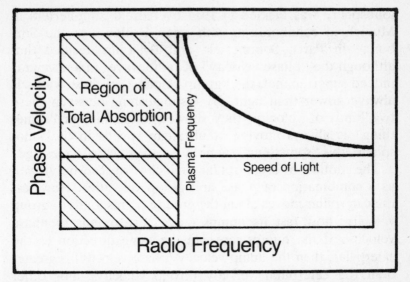

Figure 4-1. The phase velocity of a radio wave traveling through an ionized gas varies with the frequency of the wave and always exceeds the speed of light in vacuum. At high frequencies the phase velocity approaches light speed from above; at frequencies near the "plasma frequency," the phase velocity becomes infinitely large. Below the plasma frequency, the ionized gas completely absorbs radio waves; no wave can propagate in this region. This chart, showing how a wave's phase velocity changes with frequency, is an example of a "dispersion diagram."

faster than the speed of light—that is, higher than the speed of light in a vacuum. Maxwell's theory of electromagnetism shows that inside an ionized gas, not only can electromagnetic waves go faster than light, it is actually impossible for them to go slower than light!

Theory and experiment agree on how fast radio waves propagate in the ionosphere. Inside a plasma light itself (radio waves construed as low-frequency light) travels faster than light. Ironically, physicists discovered this remarkable property of waves in an ionized gas in the early part of this century, at the same time (1905) that Einstein was asserting that "velocities exceeding that of light have no possibility of existence."

The question of superluminal radio propagation in the

ionosphere was tackled in 1914 by Arnold Sommerfeld in Munich in collaboration with Leon Brillouin at the Sorbonne in Paris. Sommerfeld and Brillouin decided that although the "phase velocity" of radio waves in a plasma is indeed superluminal, the "group velocity" of these waves is always slower than light. By distinguishing between these two kinds of velocity, they showed that although "something" is always moving at ultralight speeds in the ionosphere, this "something" cannot be made to carry a message.

The motion of a caterpillar or inchworm can be viewed as a combination of phase and group velocities. How fast the caterpillar moves along the ground corresponds to group velocity; how fast its humps move corresponds to phase velocity. If the humps move in the same direction as the caterpillar, then the hump velocity (phase velocity) is greater than the caterpillar velocity (group velocity). For other caterpillars the humps may move backward, analogous to a wave whose phase velocity is smaller than its group velocity.

Sommerfeld and Brillouin stressed the importance of phase/group velocity distinction and showed that the group velocity of radio waves in the ionosphere is always slower than light. Moreover, they concluded, it is the group velocity that determines how fast you can send a signal. For instance, if you give a letter to a caterpillar to carry, it is irrelevant how fast or in what direction his humps happen to move. Hump motion is mere decoration as far as getting the message across. All that matters for message transmission is group velocity—how fast the caterpillar itself moves.

Applying Maxwell's equations to radio waves in a plasma, Sommerfeld and Brillouin discovered that the superluminal phase velocity (speed of the ripples) is always equal to the inverse of the group velocity (speed of signals). This inverse relation means that when the phase velocity is twice the speed of light, the group velocity is only half light speed; when the phase velocity is eight times light speed, the group velocity is only one-eighth of light's speed in vacuum. Since the phase velocity never goes below light speed, this inverse relationship guarantees that the group velocity will always be less than light speed at all frequencies.

To clarify the distinction between phase and group velocity, let's imagine two satellites orbiting Earth at the level of the ionosphere. The first man in space, Soviet cosmonaut Yuri Gagarin, traveled through these plasma layers in his pioneering voyage. Suppose Max is riding on a satellite 150 miles up inside a large plasma cloud. He decides to send a radio signal to Yuri the cosmonaut who is riding a similar satellite a few miles away in the same plasma cloud. Using a chart similar to that depicted in Figure 4-1, Max selects the frequency of his transmitter so that the phase velocity of the radio waves is twice the speed of light. He sends Yuri three short pulses (the letter "S" in Morse code).

Due to the inverse relation between group and phase velocity, each of these pulses travels to Yuri's receiver at one-half the speed of light in vacuum. However as Yuri looks at the pulses he notices that they are each decorated with ripples that travel at twice the speed of light. In this simple experiment Max finds that the plasma cloud, far from being a superluminal communication medium, actually transmits a pulse much slower than it would go in a vacuum, although superimposed on these slow signals are ripples that move much faster than light.

Observing that pulses go slower than light in the plasma, Max decides to switch to continuous wave (CW) transmission. He aims his antenna at Yuri's satellite and turns on his transmitter. A long radio wave reels out of his antenna like a snake with its head traveling at one-half light speed, its skin rippling at twice the speed of light. The snake's head finally reaches Yuri's satellite antenna. Now Max and Yuri are linked by a radio wave whose peaks and valleys are moving at twice the speed of light. Is Max now signaling Yuri faster than light? Not quite. Even though the satellites are connected by a carrier wave whose peaks and valleys are moving FTL, these superluminal ripples carry no message. A message must contain some sender-induced unpredictable change. But the phase wave is perfectly regular and predictable. The utter monotony of the phase wave carries no information from Max to Yuri.

Max need not be satisfied with merely sending a monotonous carrier wave, however. He can alter the carrier wave by changing the power of his transmitter in a patterned way, for instance, by modulating the carrier's amplitude with the speech signal "Dasvedanya, Yuri." Now the monotony of the phase wave is altered by a meaningful pattern. The velocity of this pattern, however, is the group wave velocity, so Max's greeting rides atop the superluminal phase wave at only half the speed of light. Although the phase waves always move at superluminal velocities, any change (whether in amplitude, frequency, or some other variable such as polarization) in these phase waves always travels at subluminal speeds. Since a message always requires some sort of a change, some variety that stands out against a monotonous background, and changes propagate through the plasma only at slower-than-light speeds, it appears as though the plasma, despite its superluminal pretensions, is not capable of being used as an FTL communication medium.

The presence of superluminal phase waves in a plasma opens a tantalizing doorway for possible FTL signaling schemes but apparently phase waves cannot be made to carry messages. Then why do such ultrafast waves exist at all? If nothing "real" can ever travel faster than light, how in fact are such superluminal waves produced inside the plasma?

If we look inside a plasma, we see two kinds of waves: the initial radio wave sent into the plasma, and a collection of secondary waves emitted from the ions as they are excited by the primary wave. The ions that constitute the plasma are charged particles, which have the ability to radiate radio waves themselves when set into motion. As it travels through the plasma, the initial radio wave causes these ions to oscillate at the same frequency as the primary wave. The ions in turn act as tiny radio transmitters. Because the ions are separated by large volumes of empty space, both the initial wave and the secondary waves travel at the same speed—the speed of light in a vacuum.

The frequency of the secondary waves is the same as the primary wave but these new waves are produced 90 degrees out of phase with the incoming wave. This 90-degree phase shift means that when the primary wave is going through zero, the secondary wave is going through either a peak (+ 90 degrees) or a valley (–90 degrees).

The phase velocity of a wave can be tracked by watching how fast its peaks and valleys move or by watching how fast its zero points move. Before the initial wave enters the plasma, its zero points move at the speed of light. When the wave moves inside the plasma, the secondary waves broadcasting out of phase add or subtract energy just at the zero point, which makes the resultant zero point occur at a different time. If this new zero occurs at an earlier time, it will appear as though this point moves faster than light. Thus the apparent superluminal phase velocity is the result of interference between two sets of waves both traveling through the plasma at the speed of light but the particular timing of these two luminal waves causes a superluminal ripple to ride across their backs.

On the other hand, if the new zero time happens to be later than the original zero time, it will appear as though the zero point is moving slower than light. This kind of zero shift is more common, so that, in glass, water, and most other transparent materials, the phase velocity is always slower than the speed of light. However in certain special materials, such as ionized gas clouds, the zero shift takes place in the other direction, leading to an FTL phase velocity.

This picture of how superluminal phase velocities arise in a plasma medium shows why Max cannot exploit this FTL velocity to send FTL signals. Any signal would have to ultimately travel from ion to ion, carried by radio waves. But both initial and secondary radio waves travel only at the speed of light between ions. So the fastest attainable signal in an ionized gas is a signal that goes exactly at the speed of light. Slower speeds are of course possible. One can imagine the ions pausing before they emit their secondary radio waves. In this case, the overall signal velocity could be slower than light because of the accumulation of many small delays. Hence in a plasma the group velocity of a radio wave is slower than the same wave in a vacuum because of the presence of the ions.

Another situation in which secondary waves interfere with a primary wave to produce a phase velocity greater than the speed of light in a vacuum is the case of the waveguide. A waveguide is nothing more than a hollow metal tube of rectangular cross-section, like a long metallic hallway. To a radio wave traveling down such a hallway, the metal walls behave like perfect mirrors. Radio waves reflecting back and forth between these walls interfere in the same way as do secondary waves in a plasma to produce a phase velocity (speed of the wave's zeros) greater than light. As in the plasma case, the waveguide's phase and group velocities are inversely related. The combination of an everywhere superluminal phase velocity plus the inverse relation between group and phase velocity guarantees that the group velocity (speed of a signal) in a waveguide is always lower than the speed of light.

After showing that the signal velocity of plasma waves is always subluminal, Sommerfeld and Brillouin went on to demonstrate that the signal velocity in any medium could never exceed Einstein's limit. Thus inside a plasma, a waveguide, and in certain other exotic circumstances, real waves may actually travel faster than light, but this superluminal motion will always be of the monotonous phase wave kind, too predictable and regular to carry a signal.

Searchlights and comet tails can travel faster than light but because nothing is ever really moving at FTL speeds in these situations, it is obvious that one could never hope to use such phenomena for superluminal signaling. On the other hand, the phase wave is an example of something real that's actually moving FTL. This superluminal loophole may be a tempting prospect for exploitation by folks in the future more ingenious than ourselves. Let's look now at one particular scheme (called Carvello's paradox) that attempts to use the very monotony of a phase wave to peer into the future.

One of science's most valuable intellectual tools is a mathematical technique called Fourier analysis. Joseph Fourier, the inventor of Fourier analysis, was a member of the group of French savants who accompanied Napoleon on his Egyptian campaign. Fourier brought back from Egypt a copy of the Rosetta Stone, which contained the key to the eventual translation of the Egyptian hieroglyphs. In a sense, just as the Rosetta Stone was the key to the language of ancient Egypt, so Fourier analysis has become the key to a more profound understanding of the nature of waves. Fourier analysis, a new and powerful language for describing wave motion, was invented by Fourier in 1823 as a by-product of his studies on heat conduction. Fourier's brilliant discovery was hailed by Lord Rayleigh, the leading British physicist of his day, as "a great mathematical poem."

Think up a wave form as complex as you can imagine. Joseph Fourier showed how to express such a wave as the sum of elemental sine waves. A sine wave is the kind of pure tone that would be produced by a perfect musical instrument that could sound and hold forever a single pitch, uncontaminated by notes higher or lower in the musical scale. The closest approximation to true sine waves is produced by lasers and electronic oscillators. Fourier discovered, in effect, that sine waves act as a natural alphabet for waves, and he showed how to use this alphabet, how to calculate for any wave form, the sine wave frequencies, amplitudes, and phases that uniquely specify that wave form.

One example of the use of Fourier analysis is the characterization of the sounds of musical instruments by means of Fourier sine waves of different frequencies. Each instrument has its own range of Fourier frequencies—its vibration recipe or Fourier spectrum—which makes a piano sound different from a harpsichord or viola, even when they are playing the same note. Knowledge of an instrument's Fourier spectrum makes it possible to synthesize an instrument electronically by putting together a bunch of sine waves according to a particular vibration recipe.

Before the age of electronic instrumentation, the ingenious German physicist and physiologist Hermann von Helmholtz analyzed the Fourier spectrums of musical instruments by playing these instruments near dozens of spherical flasks (Helmholtz resonators)—a sort of jug-band orchestra in the physics laboratory—whose resonant frequencies he had measured. The degree to which each flask was excited by the music was a measure of how strongly that frequency was represented in the instrument's Fourier spectrum. Today the analysis of musical sounds into their sine-wave alphabets is carried out electronically but the principle is exactly the same as Helmholtz's jug collection.

That a few special wave forms might be expressed as the sum of sine waves is not so surprising, but Fourier's demonstration that any wave form whatsoever can be expressed this way is truly remarkable. An elementary sine wave is a smooth oscillation that goes on forever. Yet sums of these smooth endless waves are able to represent waves that are not smooth—waves with sharp corners, for instance—or waves that are not endless—short pulses, for instance. Fourier analysis can achieve sharp-cornered waves by adding up many waves with high frequencies. The Fourier signature of a sharp percussive, sound such as the plucking of a string, must contain many high-frequency sine-wave components. To produce zones of silence using elemental waves that are never silent requires the phases of these infinite waves to be such that these sine waves destructively interfere. In the silent zones the waves completely cancel each other out.

The elemental sine wave—out of which all other waves can be constructed—is itself a wave, infinitely long with a certain precise frequency, wavelength, and phase velocity. In any medium a signal's sine-wave components travel at the phase velocity appropriate for that medium. Each signal, according to Fourier's theorem, may be regarded as made up of a sum of phase waves of different frequencies. In a plasma, for example, the signal is always slower than light but it is made up of a sum of pure sine waves—the phase waves—all of which travel faster than light. These FTL components add together in such a way as to produce a slower-than-light resultant wave. Such matter-of-fact compliance with the Einstein limit by a gang of superluminal waves is no accident.

A distinctive feature of Fourier components, the elemental sine-wave alphabet, is that they are infinitely long. In practice—music synthesis, for example—finite sine waves are necessarily used to synthesize instrumental wave forms, but strictly speaking each phase wave goes on forever without beginning or end. A pure phase wave extends in its direction of motion across all space and vibrates eternally at one unchanging frequency. A phase wave is infinite and monotonous. Its unrelieved monotony is ultimately responsible for the uselessness of these waves for FTL signaling; its technically infinite extent leads to a particular paradox.

It may seem strange that all the wave forms in the world, each of which has a beginning and an end, can be considered to be made up of elemental sine waves that are of unlimited extent. However these elemental sine waves have to be infinite because they must be able to simulate any wave form no matter how long. Also in a particular medium these elemental Fourier sine waves travel at the phase velocity of that medium. If these waves were not infinite in extent but had a beginning and end, then in a medium such as a plasma where the phase velocity is infinite, finite Fourier waves could be used to send messages. But an infinite wave, forever the same, can carry no message.

Carvello's paradox arises from taking seriously Fourier's idea that every wave form is actually made up of a sum of

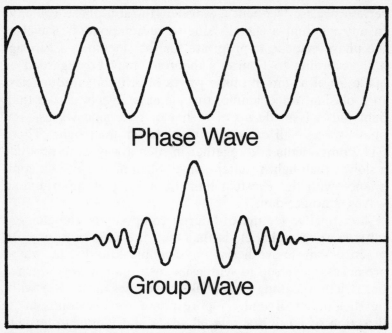

Figure 4-2. A phase wave is infinitely long, a perfectly regular sine wave. A group wave, composed according to French scientist Joseph Fourier of a sum of many phase waves, travels as a short, localized packet of energy. The sudden appearance of a group wave in an undisturbed medium can be used to send a message. Because of its infinite extent and perfect monotony, the phase wave cannot carry a message.

sine waves that are infinite in extent. (When I first heard of the "green-sunglasses paradox," it was called "Carvello's paradox" but I was never able to discover who Carvello was or why this paradox bears his name.) Imagine that the mysterious Dr. Carvello points a flashlight in your direction and turns it on precisely at midnight. At 11:59 however, a full minute before he actually turns on his light, you put on a special pair of green sunglasses and are able to see the flash of light 60 seconds before it was sent. Explaining how such anticipatory sunglasses are supposed to work is the gist of Carvello's paradox.

Considered as a wave form, the amplitude of the light is precisely zero at all times before 12:00, then it abruptly increases to a steady value where it remains until Carvello shuts it off. Expressed in Fourier's sine-wave alphabet, this light flash is made up of a sum of waves that are infinite in extent, waves without beginning or end, waves that were in existence long before midnight, and will continue to exist long afterward. To produce the flashlight's brief moment of illumination, the phases and amplitudes of these infinite waves conspire in such a way that all waves cancel for times previous to midnight. This cancellation conspiracy is broken at 12:00 and imposed again when the flashlight is extinguished. Fourier's theorem assures us that there is only one way of combining infinite sine waves to describe the short pulse of light waves produced by turning a flashlight on and off.

Your special sunglasses are designed to absorb all light except a single frequency of green. Without the other frequencies present to cancel it out, the infinitely long green sine wave will be visible behind the sunglasses since it is present in the light. As soon as you don your sunglasses you see a green flash, even before the flashlight is turned on. This flash informs you, 60 seconds before it actually happens, that Carvello intends to shine a light in your direction. With these green sunglasses, you can apparently see into the future.

Carvello's paradox brings to mind William Blake's famous line, popularized by Aldous Huxley: "If the doors of perception were cleansed, everything would appear to man as it is, infinite." Carvello's sunglasses work the other way, it seems. They operate by closing the doors of perception, except for a narrow chink, to bring the world's hidden infinities into view.

Carvello's paradox was debunked by two Dutch physicists, Hendrik Kramers, a colleague of quantum physicist Niels Bohr, and Ralph Kronig, who explained why no such "forward-looking" sunglasses could ever be built, at least out of earthly materials. In the late twenties, Kramers and Kronig considered the question: What conditions must a material satisfy in order that waves travel through it in a causal manner? (The word "causal" in this context means that no waves come out of the material before waves go in.) In other words, what properties of real materials prevent them from being used as a Carvello-style anticipatory light filter?

A piece of stained glass is colored because its light absorption, symbolized by the letter A, depends on frequency; certain frequencies of light are absorbed more than others. The light that gets through gives the glass its color. Furthermore, in glass and every other material, the phase velocity also depends on frequency; physicists call the change in phase velocity in a given material due to change in wave frequency the material's "dispersion," D. What Kramers and Kronig showed was that a material would behave in a causal manner as long as a certain relationship (now called the Kramers-Kronig [K-K] relations) existed between its absorption curve A and its dispersion curve D. If a material could be created that did not obey these relations, it would be possible, because of the infinite extent of the elemental Fourier waves, to set up situations—such as the Carvello paradox—in which an effect would precede its cause. All materials examined so far—either theoretically or in the laboratory— faithfully obey the K-K rule. The results of Kramers and Kronig have been generalized beyond the case of transparent media to impose similar conditions on any situation involving waves. Similar "causality conditions" have been derived for mechanical, electrical, and hydraulic systems as well as for systems of quantum waves. These conditions require any physical system's absorption and dispersion to work hand in hand to insure that no signal ever comes out of the system before a signal goes in.

If you made a pair of Carvello sunglasses, they could indeed filter out a band of green light, as narrow in frequency as you pleased. If the sunglasses just did that—absorbed green light—and nothing more, they would operate as claimed, and permit you literally to see into the future. But the Kramers-Kronig relations say that materials that merely absorb light do not exist. Every absorber must disperse light too, that is, it must change the light's phase velocity in a systematic way. In the case of the green sunglasses, this change in phase velocity adjusts the timing of the waves that get through in just such a way as to make them cancel completely before midnight, just as thoroughly as the full spectrum of sine waves canceled without the sunglasses.

The fact that real matter must obey the Kramers-Kronig relations blocks many other schemes to use phase waves for superluminal signaling. Consider the caterpillar analogy, which suggests that only the group velocity (bug speed) can be used for signaling, not the phase velocity (hump speed).

The hump speed does indeed seem irrelevant as long as the caterpillar is short compared to the distance he has to travel. But suppose we imagine a very long caterpillar that stretches all the way from the source of the signal to the destination, corresponding to a medium not pulsed but completely filled with phase waves. Could we now use the phase velocity (hump speed) to send messages? The Kramers-Kronig relations rule out this possibility. In any real medium, the absorption A and dispersion D will always conspire to keep real signals from exceeding the light barrier.

Ultimately it is the monotony of phase waves that prevents them from carrying signals. Although a phase wave may under certain circumstances travel faster than light, such a wave looks exactly the same everywhere and everywhen. A message, on the other hand, involves a change, some difference or distinction that is carried from place to place. But a pure phase wave is utterly changeless. If we disturb a pure phase wave we get a group wave—a Fourier superposition of pure phase waves—which must propagate at light speed or less, at least in any medium that obeys the Kramers-Kronig relations.

One might hope to hitch a ride on a phase wave by some

subtle labeling process—painting one of its humps a pale green, for instance—but no one has yet discovered how to break the everlasting monotony of a pure Fourier wave. Phase waves are single-frequency sine waves infinite in extent, which under certain conditions can actually travel faster than light. But until their immaculate monotony can be tainted with a marker that travels at this same velocity, phase waves seem destined to remain tantalizing but ultimately useless channels for superluminal signaling.

CHAPTER 5

Advanced Waves: Oscillations That Travel Backward in Time

Backward time isn't such a new thing, backward time will start long ago.
—I. J. Good, attributed to Doog, J. I., 1965

No one has advanced our understanding of the nature of light more than Scottish physicist James Clerk Maxwell. For it was Maxwell who more than a hundred years ago first discovered that light consists of vibrations of electric and magnetic fields. In 1861, Maxwell, a theoretical physicist at Kings College, London, was attempting to use a few equations to summarize everything that his predecessors had learned about electricity and magnetism. The equations Maxwell derived exhibited a pleasing symmetry between electricity and magnetism with one intriguing exception. Although a changing magnetic field could produce an electric field (the magnetic induction effect, first measured by Michael Faraday), a changing electric field produced no magnetic effect. For reasons that were essentially esthetic—to make the math look more symmetrical—Maxwell stuck into the equations an extra term that described how a changing electric field would produce a magnetic field (electric induction) hoping that this new effect might someday be observed when more sensitive measuring devices became available.

Maxwell's new electric induction postulate had more profound consequences than the mere discovery of an obscure new electromagnetic phenomenon. Maxwell's equations implied that a changing electric field gives rise to a changing magnetic field which recreates by magnetic induction the original electric field starting the cycle anew. The result of this seesaw pattern of mutual electric and magnetic induction is the production of a self-sustaining electromagnetic vibration moving though space. Maxwell found that the speed of these self-sustaining vibrations depends only on the ratio of the strength of the electric force between two charges to the magnetic force between two magnets. Since Maxwell knew the strength of both these forces, he could easily compute the velocity of this new electromagnetic vibration. To his surprise, the calculated velocity of these vibrations turned out to be precisely equal to the measured speed of light.

On the basis of this numerical agreement between the known velocity of light and the calculated velocity of electromagnetic oscillations, Maxwell proposed that light actually consists of an electromagnetic field vibrating in a certain frequency range. He predicted the discovery of similar electromagnetic vibrations at lower frequencies (radio and infrared waves) and at higher frequencies (x-rays and ultraviolet waves) than the frequencies of visible light. Maxwell's predictions were soon confirmed by the discovery of these new forms of radiation. The Scottish physicist's insight that light is an electromagnetic wave was the high-water mark of nineteenth century physics. Who before Maxwell would have guessed that the colorful sensations that delight our eyes are deeply akin to the forces dwelling inside storage batteries and bar magnets?

Maxwell's prediction from theory alone of the electromagnetic nature of light is an example of what Nobel laureate Eugene Wigner has called "the unreasonable effectiveness of mathematics in the natural sciences." None of the experiments on which Maxwell based his equations had anything to do with waves or with exceedingly rapid velocities; all of them could be done on a laboratory bench with batteries, coils of wire, and static electricity machines. But the equations Maxwell deduced from these experiments express more general truths than the facts embodied in the experiments themselves. When a science has reached this stage of mathematization it is often the case that you can get immensely more out of the equations than you put into them.

A more recent example of the fertility of applied mathematics is the case of Dirac's equation for the electron. When Paul Dirac solved this equation he got not one solution but two. Taking the extra solution seriously, he predicted the existence of a new particle—the antielectron, or positron—with charge and spin values opposite to the ordinary electron. Dirac's prediction of antimatter took more courage in 1931 when only three elementary particles were known—electron, proton, and photon—compared with today's acknowledgement of a wild proliferation of elementary subatomic entities.

Like Dirac's equation, Maxwell's wave equation for light also has two solutions, the so-called "retarded solution" that describes a wave traveling forward in time and the "advanced solution" that describes a light wave traveling backward in time. Both of these waves travel at the same speed—the speed of light in vacuum—but in opposite temporal directions. The retarded wave travels in the normal direction—from past to future—while the advanced wave goes the other way—from the future into the past.

Imagine that in the year 2000 Maxine says goodbye to her brother Max and travels on a NAFAL spaceship to Betelgeuse, the bright red star in Orion's left shoulder. Since Betelgeuse—which means "giant's armpit" in fractured Arabic—is 500 light-years from Earth, Maxine's ship will take approximately 500 Earth years to reach this big red star. By that time (2500 A. D.) Max has been long dead and his sixteenth-generation descendant Max XVI is now head of the family. However, because of relativistic time dilation, Maxine has only aged a few hours. What are the communication options open to Maxine if she is restricted to using only ordinary (retarded) light?

As Maxine looks wistfully back to Earth, she sees it as it was 500 years in the past, because it takes light 500 years to travel from Earth to Betelgeuse. Maxine sees the Earth she remembers, and in fact, through a sophisticated Betelgeusian telescope she can actually watch her brother Max I, caught in a traffic jam after leaving the spaceport. However if Maxine were to send a telegram back to Earth with ordinary light or radio waves, news of her safe arrival would take 500 years to get back to Earth. Her message would be read by Max XXXIII in 3000 A. D., approximately 1000 years after Maxine's departure. Messages sent via the retarded waves characteristic of ordinary light always arrive late, this delay is especially noticeable across astronomical distances. If the Betelgeusians have the ability to produce advanced waves—Maxwell equation solutions of the second kind—Maxine would not be so isolated from her home planet. A message sent by advanced-wave telegraph leaving Betelgeuse in the year 2500 would travel backward in time at the speed of light—at minus 186,000 mps, so to speak— and would reach Earth 500 years before it was sent, arriving at Maxine's home planet in the year 2000, the same year she left. Thanks to the Betelgeusian advanced-wave communicator, Maxine could let her brother Max know that she had arrived safely even though Max has really been dead for hundreds of years.

Ordinary light waves are called "retarded" because you always receive them after they are sent; advanced waves, on the other hand, are always received before they are sent. Both waves travel at the speed of light. Although advanced waves are permitted by Maxwell's equations, no advanced waves have ever shown up in any experiment. All light waves that we know about seem to be of the retarded variety.

For more than a hundred years physicists have wondered about how to interpret the mysterious advanced-wave solution to Maxwell's equations. Because of the lack of experimental evidence, most are content to ignore the advanced solution, dismissing it as an option that nature simply chose not to exercise. Other scientists have toyed with the idea that advanced waves really exist but we cannot see these backward waves because of certain peculiar circumstances.

The enigmatic status of the advanced solutions to Maxwell's equations is part of a larger question in physics known as the "arrow of time" puzzle. The basic laws of physics, which include relativistic mechanics, Maxwell's equations, and quantum theory, do not favor one direction of space over another—to go west is no more difficult or unusual than to go east as far as these equations are concerned. Similarly the basic laws of physics do not single out a particular direction in time: For a process to be pushed from the future to the past is just as permissible in these rules for the world as to be pushed from the past to the future.

If we take the temporal indifference of these physics equations at face value, we would have to treat the difference between past and future as mere convention, like the difference between east and west. Yet ordinary experience tells us that the future is profoundly different from the past. For some unexplained reason, these time-symmetric laws have not resulted in a time-symmetric world, but have generated instead a world containing an "arrow of time" that makes one time "direction" different from the other.

Actually there is not just one time arrow problem to be explained, at least six such mysterious arrows are embedded in natural phenomena. Besides the electromagnetic arrow, the time-asymmetry problem includes the psychological arrow, the thermodynamical arrow, the cosmological arrow, the quantum-mechanical arrow, and the weak-interaction arrow. (British physicist Paul Davies gives a nice account of these various temporal biases in his book *The Physics of Time Asymmetry*.)

The psychological arrow of time consists of the fact that in our private mental experience, the past and the future play distinctly different roles. We remember the past, and anticipate the future. Science-fiction writers have explored the consequences of reversing this mental arrow. In stories of this type—J. G. Ballard's "Time of Passage" or Roger Zelazny's "Divine Madness," for instance—the hero rises out of the ground, greeted by relatives absorbing water into their eyes, and proceeds to grow younger day by day. In the time-reversed workplace, autos are disassembled into raw materials which are returned to the earth, schools are factories for forgetting, and life inevitably ends with a return to the womb. Another literary example of a reversed psychological arrow is T. H. White's Merlin in *The Once and Future King* who remembers the future but not the past.

Since we know so little about the physical origins of consciousness, only wild speculations are possible about why human awareness inevitably moves from the past into the future, never in the other direction. The psychological arrow is probably connected with the thermodynamical arrow.

The stubborn physical fact that entropy—a measure of molecular disorder—always increases as we move from past to future indicates the presence of a thermodynamical arrow of time in nature. Without a continual input of organized energy, every known system tends to become more and more disorganized as time passes. Apparent exceptions to this rule of inevitable entropy increase, such as living systems that appear to build up order, have achieved their orderliness at the expense of a much greater increase in the disorder of their energy sources. In the case of human beings, the Sun's vast entropy increase as it radiates energy into space more than compensates for the slight antientropic accomplishments of Earth's biosphere. When their essential support systems are taken into account, living creatures and all other systems made of molecules seem, inevitably, to evolve toward a future that possesses more disorder than the present. Yet the molecules themselves obey symmetric laws of motion—at the molecular level there is no distinction between past and future, no one-way street restricting

handwritten margin note: Dick, P.K. Counter-Clock World

molecular motion. The origin of the thermodynamical arrow is sometimes blamed on the cosmological arrow of time.

A cosmological arrow of time can be discerned in the overall behavior of the Universe. As it moves from past to future, the Universe is expanding (rather than contracting or staying the same size). The future is distinguished from the past by the fact that in the future the Universe is more capacious. The birth of our Universe according to present thinking was distinguished by a unique event—the Big Bang, the actual beginning of time. Because of this special creation event, time itself is not symmetric: The past direction is singled out; the past is defined by the arrow that points in the temporal direction of the Big Bang. For reasons as yet unfathomable, the Big Bang was an event of exceedingly high thermodynamical order—a very improbable event—and that initial excess order has been steadily decreasing ever since, an inevitable increase in messiness that physicists call the thermodynamical arrow of time. Some theoreticians have speculated that if the Universe were to enter a collapse phase—as it must do if it is closed—some of the other arrows might also reverse. Heat would flow (in violation of entropy increase) from cold to hot bodies, cars would attract paint out of the air rather than rust, and order everywhere would arise spontaneously out of chaos. Curiously, if our psychological arrow reversed as well, we might never notice the world's new backward behavior.

In contrast to classical physics, whose deterministic laws permit only one possible consequence to evolve from a given set of circumstances, quantum theory describes a world in which many outcomes are possible. Quantum theory represents the motions of atoms, electrons, and quarks as probability waves. As with all the other equations of physics, the equations that these quantum waves obey are time-symmetric, the evolution of quantum probabilities makes no distinction between the past and future time directions.

Unlike the classical world, which was described in only one way whether it was observed or not, the quantum world is described in two ways: The unobserved world takes the form of probabilities and the observed world is made up of facts. During the act of observation itself many probabilities change suddenly into one fact, the result of a still incompletely understood quantum process called the "wave function collapse." This process, in which many past probabilities turn into one future fact—never the other way around—establishes preferred temporal direction called the quantum-mechanical arrow of time.

The weak-interaction arrow of time arises from the experimental fact that a particular weak interaction—the decay of the neutral K-particle, or "K-zero" for short, into a pair of pions—possesses an intrinsic time direction. If the decay of the K-zero into two pions were time-symmetric, the reverse process—production of a K-zero when two pions collide—should look exactly like a time-reversed movie of the decay process. This reverse process is actually very difficult to measure but measurements made on more accessible reactions indicate that two-pion K-zero production has a different amplitude and phase from the time-reversed movie of K-zero decay. The two reactions differ only by a few parts per thousand but this small difference has been confirmed repeatedly by several laboratories around the world and is now accepted as an undeniable fact of nature. Alone among the equations of physics, the law that governs K-particle decay is intrinsically time-asymmetric to a very small extent, a tiny flyspeck in the Universe's otherwise perfect temporal symmetry, the handiwork perhaps of some cosmic prankster. This weak-interaction arrow came totally out of the blue 30 years ago and physicists still have no idea what to make of it.

In 1945, John Wheeler and his graduate student Richard Feynman, then at Princeton University, proposed a novel way of looking at light that gives the backward-in-time solutions of Maxwell's equations equal status with the forward-in-time solutions. John Wheeler, author of a seven-pound paperback book on gravity, is presently director of a fundamental physics research center in Austin, Texas. Richard Feynman, who received the Nobel prize for developing an improved quantum theory of electromagnetic radiation, is remembered in some circles as "the Groucho Marx of physics" for his humorous and irreverent behavior. The Wheeler-Feynman model, called the "absorber theory of radiation," makes electromagnetism a two-way street as far as the time dimension is concerned. They base their time-symmetric theory on the assumption that every light wave emitted by an atom must be somewhere absorbed by another atom and that these two events, light emission plus light absorption, should be considered as a single insepara-ble process.

In conventional radiation theory, an atom emits a wave of light without regard to the light's eventual absorption. As the light wave leaves its present atom, traveling in a particular direction, the atom recoils in the opposite direction. The recoil of an atom as it emits light is no mere theoretical construct but has actually been observed in certain atomic beam experiments. Conventional radiation theory explains this atomic recoil as the reaction of the emitted wave back on the atom. As you jump from a boat onto the dock, the boat will recoil for a similar reason, responding to your backward push. Wheeler and Feynman proposed that the emitted wave has no effect whatsoever on the atom. In their model, the atom's recoil is caused by a light wave that travels backward in time from the atom that eventually absorbs the light.

The essential point of the Wheeler-Feynman model is that atoms don't interact with their own radiation; they only interact with the radiation from other atoms. The cause of an atom's recoil cannot lie in its own electromagnetic field but must be due to the field of some other atom.

Since the early days of electromagnetism physicists have struggled with the notion of self-interactions. Do the electric and magnetic fields produced by an atom act back on the atom itself? The problem with self-interaction is that in most reasonable models of the atom, when you calculate the self-interaction energy, it always turns out to be infinite. According to Einstein's equation, $E = mc^2$, an infinite energy is equivalent to an infinite mass, so the theoretical mass of atoms should be infinite. Most physicists accept the existence of self-interaction despite this theoretically infinite energy hoping that someday a more sophisticated calculation will be able to reduce the self-interaction energy to a finite value. Wheeler and Feynman solve the problem of infinite atomic energy by simply not putting self-interaction into their theory from the start. Instead they replace self-interaction with a perhaps more improbable concept, a wave of light that travels backward in time.

In conventional radiation theory, light from atom A travels at a speed of 186,000 mps to atom B which passively absorbs the light. Atom A recoils because the emitted light pushes back on it (self-interaction) as the light escapes from its parent atom. Atom B recoils because it absorbs the incoming light wave's momentum, like a baseball player being pushed back as he catches a fast ball. According to the Wheeler-Feynman theory, the situation is more complicated—radiation occurs in two steps. First atom A emits, without recoiling (no self-interaction), a half-sized (retarded) wave that travels forward in time at a speed of 186,000 mps to the absorber atom B. Atom B recoils as it takes up this light's momentum. Then, stimulated by its recoil motion, atom B emits a half-sized (advanced) wave that travels backward in time at a speed of 186,000 mps to atom A. Atom A recoils as it takes up this advanced wave's momentum.

The timing of the emission and absorption events guarantees that at any moment the half-sized advanced wave sent back in time from the absorber always finds itself at the same position in space as the half-sized retarded wave sent forward in time from the emitter. Thus the two waves fuse together to form a single full-sized wave which appears to

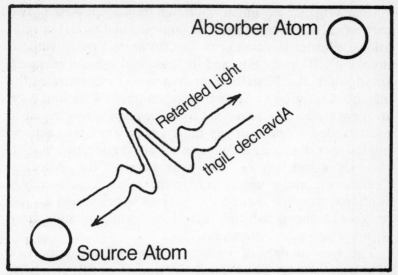

Figure 5-1. All absorber theories agree concerning what happens when light from a source atom is eventually absorbed by some other atom in the future. The source sends into the future a half-size retarded wave that travels to the absorber at light speed. The absorber responds by emitting into the past a light-speed, half-size advanced wave that reaches the source at the same time as the initial wave begins its journey. Because the advanced and retarded waves coincide at all times, the resultant wave looks like a full-size retarded wave traveling from past to future.

have been sent from the emitter and received by the absorber. Because of this exact superposition of advanced and retarded waves, the Wheeler-Feynman model produces the same apparent wave motion as conventional radiation theory. The two theories give the same result but propose radically different models of what is actually happening. Conventional radiation theory is a simple matter of cause and effect. Wheeler-Feynman theory involves a handshaking procedure much like data exchanged between two computers in which a data transfer initiated by one computer is not completed until the exchange has been acknowledged by a message sent back from the second computer.

*   *   *

The Wheeler-Feynman radiation model possesses two theoretical advantages over the conventional model of radiation. Wheeler-Feynman can explain atomic recoil without invoking self-interaction and its attendant infinite energies. In addition, the Wheeler-Feynman model incorporates the advanced solution to Maxwell's equations in a natural way. In their model this advanced solution plays a role equal to the retarded solution in any radiation process, thus achieving the sort of symmetry physicists find so attractive in their theories about nature. However because the Wheeler-Feynman model predicts exactly the same results as conventional radiation theory, there seems to be, at first glance, no way to distinguish this radical model of radiation from its more commonsense competitor.

The two models of radiation differ in one important respect, however. The Wheeler-Feynman model, otherwise known as the absorber theory, is not only testable but, if this theory is correct, certain schemes to send signals backward in time become possible. According to conventional radiation theory, the emitted light may or may not be absorbed, but in the absorber theory, in order for light to be emitted it must be connected to some future absorber by a two-way retarded-advanced wave handshake process. Because of the need for the presence of absorbers, the absorber theory predicts that if there are none in a particular direction in space, then light will refuse to shine in that direction! If absorber theory is correct, your flashlight would go out whenever you shine it up into the night sky in the direction of an "antiemission locus"—a region of space entirely devoid of absorbers of light in the frequency range of your flashlight. In certain other directions, containing only a few absorber atoms, your flashlight would dim but not go out. Only in those directions in which light was totally absorbed would the flashlight be able to shine at its full brightness.

The presence of advanced waves in absorber theory does not by itself permit backward-in-time signaling because in the Wheeler-Feynman picture, these advanced waves never occur alone but only as part of a two-way handshake process involving ordinary retarded radiation. However the existence of antiemission loci—"dead zones in space" into which light refuses to shine—makes possible certain backward-in-time communication schemes. To build an advanced-wave communicator, you need only come up with a receiver that absorbs light more strongly than empty space itself, made of a kind of material that is "ultrablack." If antiemission loci actually exist at certain places in the sky, then an ordinary complete absorber held in front of these dead zones could be used as an advanced-wave communicator.

In an ordinary radio station, the transmitter sends out energy the receiver detects. Advanced communicators work in the opposite way. Energy is sent out from the transmitter at maximum power but instead of looking at the receiver, you monitor the transmitter's output. At the receiving end, you pay no attention to the received energy, but instead send a message by placing an ultrablack absorber in front of the antenna. When this absorber is inserted, it causes the transmitter to send more power. The presence or absence of the ultrablack absorber at the receiver is detected at the transmitter by an observed increase in the amount of output power. The influence of the receiver on the transmitter occurs at the speed of light but in a backward temporal direction. If the receiver is 500 light-years from the transmitter, the transmitter will experience a power surge 500 years before the black absorber is placed in front of the receiver's antenna.

Such backward-in-time signaling leads to no paradoxes as long as the signaling remains one-way, but if two transmitter-receiver pairs were set up, then you could send messages back into your own past, not merely the past of someone 500 light-years away, with decidedly paradoxical consequences. Thus if absorber theory is the correct model for the way radiation really works, and if, for certain frequencies of light, absorbers are very sparse in certain directions in space, then a simple time machine could be built out of ordinary radio transmitters and a few big black billboards.

One problem with testing absorber theory on Earth is that the atmosphere itself is a very good absorber for most frequencies of light. An Earthling's view of outer space is restricted to two narrow spectral windows: the atmosphere is transparent to visible light allowing optical telescopes to see the stars, and also transparent to microwaves for the benefit of radio-telescopes. The first (and only) test to date of Wheeler-Feynman absorber theory was carried out in the microwave region of the electromagnetic spectrum. In 1973 Robert Partridge, at Hartford College in Pennsylvania, attempted to look for the antiemission loci permitted by Wheeler-Feynman theory by measuring the power output of a microwave antenna as it was directed toward a perfect microwave absorber on the ground compared to the output of the same antenna aimed directly overhead into empty space. Partridge found no difference between antenna input power in these two situations, indicating either that light (at microwave frequencies) is completely absorbed in space or that the Wheeler-Feynman theory is wrong. At present we do not have enough information to choose between these alternatives.

Wheeler and Feynman originally developed absorber theory to explain the emission of light waves. But since modern quantum theory represents all particles by waves—the so called "Schrödinger wave function"—it is possible to develop a Wheeler-Feynman type of theory for the absorption and emission of particles as well. Such a theory, in which every particle emitted by some source is ultimately absorbed somewhere else and where every retarded wave is accompanied by an advanced wave of equal amplitude traveling in the opposite temporal direction, has been developed by Paul Csonka at the University of Oregon and John Cramer at the University of Washington. Just as Feynman and Wheeler used their theory to resolve some of the paradoxes of classical radiation theory, so Csonka and Cramer use their advanced-wave theory of particles to resolve certain interpretational questions in quantum theory.

However there are some important differences between light and particles that make the Csonka-Cramer theory of particular interest to backward-in-time communications buffs.

Consider, for example, the particle physicists call the "neutrino." This particle interacts so weakly with matter that if the Universe were made entirely of lead a neutrino would still stand a good chance of traveling its full length without being absorbed. In our neutrino-transparent Universe, most neutrinos are destined never to be absorbed.

Because our universe is so transparent to neutrinos, if a civilization in the Andromeda Galaxy—2 million light-years from Earth—invented an efficient neutrino absorber, we would observe—if the Csonka-Cramer theory is correct—in all reactions that emit neutrinos, such as the reactions that power the Sun, an enhanced emission of neutrinos in the direction of Andromeda. Furthermore, as the Andromedans erect or withdraw their absorber, like Indians sending signals across the plains by waving a blanket over a smoky fire, we would observe a modulation of neutrino emission on Earth. Unlike an ordinary light wave, which reaches its destination at a later time than it was emitted, this Andromedan "smoke signal" is carried by advanced waves and arrives on Earth 2 million years before it was sent.

If upon reception of this advanced Andromedan message, we immediately sent the Andromedans a reply by conventional retarded light—a tight laser beam, for instance—this return message would be delayed by 2 million years and would arrive at Andromeda shortly after the aliens sent their advanced signal. Thus between two civilizations—one of which has an advanced transmitter and the other a retarded transmitter—real-time, no-lag communication is possible.

According to the Wheeler-Feynman theory, every light-emitting event involves both a retarded and advanced wave, but no free advanced waves are ever observed because emitter and absorber waves always act in concert, never alone. The only unconventional effect predicted by Wheeler-

Feynman theory is the suppression of radiation in directions where future light absorption is not complete—that is, directions with a shortage of absorbers. It is this antiemission effect that could form the basis of a system for signaling backward into the past.

A version of the Wheeler-Feynman theory proposed by Paul Csonka eliminates antiemission loci but introduces explicit advanced effects instead. Csonka suggests that particles that are not completely absorbed in the future (neutrinos, for example) do not cease to emit, but their radiation consists of equal parts retarded and advanced waves. The reason that advanced waves from such dual emissions have never been observed, according to Csonka, is that they have never been systematically searched for and would be difficult to see. The very reason that these advanced waves are produced—their weak interaction with the rest of the Universe—makes their presence difficult to detect. The advanced waves associated with these particles likewise weakly interact with any detecting device.

In Csonka's view, the more strongly a particle interacts with the Universe, the more that particle participates in the conventional one-way flow of time. Particles that have a vanishing chance of interacting in the future, in Csonka's words, "do not see the Universe very well." Such particles "lose their way in time" and cannot distinguish the past direction from the future. These chronologically confused particles will be emitted in the form of equal amplitude retarded and advanced waves—half going toward the future, half toward the past. When we observe one of these "lost in time" particles it is equally likely to have come from the future as from the past. Strongly interacting particles, on the other hand, have a strong sense of time direction. Because the advanced waves associated with strongly interacting particles only occur as part of handshaking interactions with future absorbers, particles of the strongly interacting sort, just like completely absorbed Wheeler-Feynman light, appear to come only out of the past.

Since for human observers, time seems to run in only one

direction, the psychological arrow of time, a particle traveling backward in time superficially looks exactly like a particle traveling forward in time. Only the particle's causal structure (its cause lies in the future not the past) can reveal the time-reversed nature of a particle traveling backward in time.

To observe this distinctive causal structure, Csonka suggests that we look at reactions that emit high-energy neutrinos, such as the decay of fast pions in a high-energy accelerator. If Csonka's theory is correct we will observe a most unlikely occurrence in some of these reactions. From some random direction a high-energy neutrino will suddenly appear, and like a sharpshooter's bullet, will strike and be absorbed by the neutrino source precisely at the moment when it is emitting a normal neutrino. The incoming neutrino is so accurately aimed because it is really emitted by the source as an advanced wave traveling into the past, but perceived by our one-way minds as if it were traveling from the past into the future.

Although this neutrino seems to come from the past, it is really an advanced-wave neutrino sent into the past by the source. Csonka calls these accurately aimed neutrinos "shadow neutrinos" because their appearance is due to the lack of a canceling signal from the surroundings, just as in everyday life a shadow arises due to lack of light that would fill it in. Ordinary neutrinos are difficult to detect let alone the elusive shadow neutrinos. Neutrino detectors are gigantic (consisting of millions of gallons of cleaning fluid, monitored by hundreds of light detectors) and insensitive (a few neutrinos detected each day). No one has taken Csonka's theory seriously enough to mount a search for shadow neutrinos, nor have any effects that might be attributed to temporally backward particles been seen in conventional neutrino experiments.

Although there are no antiemission loci in Csonka's theory, shadow particles could be used for backward-in-time communication schemes in much the same manner as in the Wheeler-Feynman theory. In Csonka's scheme, modulation of a future absorber changes the ratio of advanced to retarded waves at a source in the past. Erecting an absorber in the future backwardly causes the number of shadow particles observed in the present to decrease—a kind of a shadow of a shadow cast by an object in the future! In the Wheeler-Feynman case, the source emits more particles when a future absorber is present; in the Csonka variation, the source absorbs fewer shadow particles. Although the effects to be looked for are different, both theories offer the possibility of sending a signal backward in time at the speed of light.

At the University of Washington in Seattle, John Cramer came up with another version of Wheeler-Feynman theory. As in Csonka's theory, a source that is not matched up with a future absorber emits two waves: an advanced wave into the past and a retarded wave into the future. In Csonka's theory, the advanced wave should be observable but in Cramer's version this advanced wave is suppressed by reflection off the Big Bang. To eliminate this unwanted advanced wave Cramer casts the Big Bang in the role of "absorber of last resort." He assumes that the situation at or near this singular creation event was such as to reflect advanced waves arriving from the future back into the future with a 180 degree phase change. This reflected wave— now a retarded wave traveling into the future—cancels out the source's advanced wave and reinforces the source's retarded wave resulting in no observable advanced waves. Because of the Big Bang reflection, nothing is left of the time-symmetric light but the familiar retarded wave going in the usual direction from past to future. According to Cramer, light is emitted into the future rather than the past "for the same reason that the light from a spotlight points in a particular spatial direction: both have a reflector 'behind' them which reflects all rays going the wrong way."

Cramer's "Big Bang as Big Mirror" assumption for both particles and radiation eliminates the possibility of using a modulated future absorber for sending messages into the

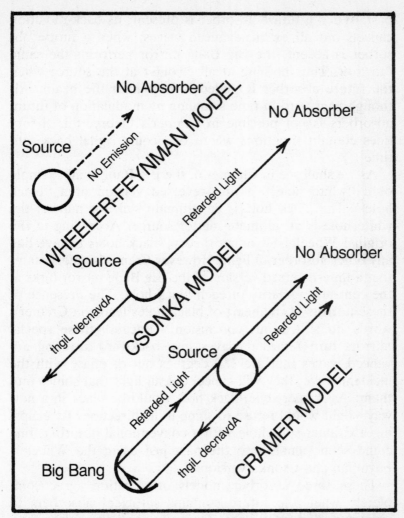

Figure 5-2. Three absorber models of radiation are illustrated. Absorber theories differ from one another by how they describe the situation in which no absorber is present in the direction of the emitted light. When an absorber is absent, the Wheeler-Feynman model predicts that no radiation will be emitted in that direction. Paul Csonka's version of absorber theory predicts that a pair of half-size waves is emitted by the source, one wave going forward, the other going backward in time. John Cramer's absorber model likewise predicts a pair of waves emanating in opposite temporal directions from the source, but in Cramer's model an out-of-phase reflection from the Big Bang cancels out the backward advanced wave and strengthens the forward-in-time retarded wave.

past. When a future absorber is present, its back-radiation cancels out all extra advanced waves; when a future absorber is absent, the Big Bang mirror performs the same function. Thus nothing at all changes at the source when the future absorber is pulled in or out of the beam. Although backward-in-time signaling by modulation of future absorbers is not possible in Cramer's theory, this theory does contain situations where one could signal back into time.

As we shall see in Chapter 6, the Big Bang is an example of a "white hole," a time-reversed version of a "black hole." The black hole is an ultimate sink for matter; the white hole is an ultimate matter source. According to the original Wheeler-Feynman theory, black holes behave like any other absorber of light. However Cramer's theory states that a time-reversed version of the Big Bang mirror lurks at the central singularity of each black hole. The presence of these mirrors at the heart of black holes leads, in Cramer's words, to a "curious conclusion." Because these special mirrors turn retarded waves into backward-traveling advanced waves that are 180 degrees out of phase with the incident wave, they will cancel out all light that shines into them. A Cramer-style black hole would be black in a new way—light would refuse to shine in its direction. By eclipsing a Cramer-style hole with a conventional absorber, one could send signals into the past just as in the Wheeler-Feynman and Csonka versions.

These three absorber models of radiation agree completely when the future contains sufficient absorbers to intercept all emitted light. These theories differ, however, in their description of what happens when future absorbers are scarce. Figure 5-2 illustrates the major differences among these three unconventional radiation theories when no absorber is present to intercept the emitted radiation.

What are the traffic laws that tell a particle or light wave how to get from one place to another? Conventional radiation theory imposes a speed limit (186,000 mps) and a forbidden temporal direction (no signals into the past).

Various deviant radiation theories have been proposed by Wheeler, Feynman, and others which retain the universal speed limit but abolish the one-way traffic rule in favor of other mechanisms that still preserve the appearance of one-way radiation when observed under ordinary conditions. Because of certain special properties of emitters and absorbers, these time-symmetric radiation theories manage to agree for the most part with the manifestly time-asymmetric nature of our experience with light.

However in special situations (lack of future absorbers, eclipsing black holes) absorber theories not only give results at variance with conventional radiation theory, but actually predict the existence of raw uncanceled advanced waves, which travel into the past in opposition to the general rush of radiation into the future. Moreover, such backward radiation can, in principle, be modulated to send signals into the past. Although such signals never travel faster than light, they do go back in time, and give rise to the possibility of the same causal paradoxes as FTL signaling.

No experimental evidence has ever been found for advanced waves; perhaps the Universe really is a one-way street for both particles and radiation. Yet the fundamental equations of physics permit waves to travel in both directions. The notion of advanced waves remains a loophole through which a future physics of backward causality might someday sneak.

CHAPTER 6

Space Warps: Shortcuts in Curved Space-Time

Anyway, there I was in a flexible nauton shell being sucked down into the heart of Gouda X-1 at something like the speed of light. . . . The singularity was dead ahead now, a bright ring a few kilometers in diameter. . . . I was thin as a needle and five kilometers long when I hit the center of the ring, and I whisked through without even slowing down. But as soon as I'd gone through I was in an anti-universe, and every particle of that universe wanted me out of there. . . . When you come back through the ring . . . if you're lucky you come out where and when you wanted to.
—Rudy Rucker, "Jumping Jack Flash," 1983

In Rudy Rucker's "Jumping Jack Flash," Siborg, a shape-changing alien from a parallel universe, travels to black hole Gouda X-1 via a flatulence-based propulsion system. Diving through the black hole's ring singularity Siborg emerges in our Universe where he masquerades as an English professor at Cornell University. To repair a mishap suffered during a previous visit, Siborg plans his black hole transit so he effectively travels backward in time, arriving moments before making his big mistake.

In the late sixties, the black hole as gateway to other universes became a science-fiction staple—a standard plot device to overcome inconveniently large interstellar distances or to reverse the seemingly inevitable one-way flow of time. Can black holes really function as stargates or time tunnels? Or is the black hole as rapid transit a mere literary device outside the realm of real-world possibility?

A black hole is a region of space-time so strongly deformed by gravity that no light can escape from it. Black holes are one consequence of general relativity—Einstein's theory of curved space-time. Before the advent of relativity, however, the French mathematician Simon Laplace, in 1795, conceived of a black-hole-like object that was a consequence of Newtonian mechanics. The escape velocity of a projectile shot from a planet's surface depends only on the planet's mass and radius. Laplace imagined a planet so massive that its escape velocity was greater than the speed of light. Since no light could leave its surface, such a planet would appear totally black. Using Newton's gravitational laws, Laplace calculated the radius of this black planet. Although Einstein showed that Newton's laws are inadequate to describe the gravitational effect of such extremely massive bodies, by a remarkable coincidence not uncommon in physics—where a calculation using wrong assumptions gives the right result—Laplace's calculation for the critical radius of a dark planet gives the same answer as Karl Schwarzschild's spherically symmetric solution of Einstein's equations of general relativity.

Einstein developed the general theory of relativity as an extension of his special theory by taking seriously a metaphor invented by his former teacher, Hermann Minkowski. For Einstein, special relativity consisted primarily of the set of relations between space and time that must result if one requires that the laws of physics remain the same for all moving observers. These relations—called the Lorentz transformations—express the apparent distortions in length, time, and mass, that a stationary observer will attribute to a moving observer's measurements.

Minkowski saw the Lorentz transformations in a new light. He noticed that when the laws of physics were expressed in the Einsteinian way, the laws themselves resembled mathematical "objects" in a four-dimensional space—a space consisting of the usual dimensions of space plus time as the additional fourth dimension.

If you rotate a circular hoop in three dimensions it appears to the eye as a shape-changing ellipse. Only in one position does it look like a circle. Behind all these separate views, however, lies one invariant object, which produces—via the laws of perspective—the different apparently elliptical images. According to Minkowski, special relativity functions as a kind of law of perspective in four dimensions. Relative motion between two observers is equivalent to a rotation in four-dimensional space. The individual measurements that go into the physical laws—clock rates, lengths, electrical field strengths—all change when an observer changes his state of motion. But these mutable measurements can be arranged, as Einstein showed, to form four-dimensional objects called "tensors" whose form does not change when observers move. If you want to insure that the laws of physics are invariant for all observers, then you must formulate these laws as relations between tensors, not as relations between mutable variables such as space, time, and mass.

Minkowski's new idea was that special relativity was more than a set of mathematical relations. What relativity really means is that we are living in a four-dimensional world. "Henceforth space by itself and time by itself are doomed to fade away into mere shadows," Minkowski declared, "and only a union of the two will preserve an independent reality."

At first Einstein resisted Minkowski's interpretation of special relativity. He might have objected because time is not treated in the same manner as space in his theory, hence the four "dimensions" cannot really be considered equivalent. This difference is expressed mathematically by time's negative "metric signature"—the fact that the square of a timelike interval is always negative, while the square of a spacelike interval is always positive. Thus time could hardly be considered a dimension on an equal basis with space. A four-dimensional space in which all dimensions have the same metric signature is called a "Euclidean space"; this new space, in which one dimension has a special signature, is called "Minkowski space." If the concept of "dimensions" is slightly generalized to accommodate this special feature of time we end up with Minkowskian "space-time," the purported arena for all the laws of physics.

After some hesitation, Einstein accepted and expanded Minkowski's notion of space-time by imagining what would happen if this four-dimensional space-time fabric could be curved. In his general theory of relativity Einstein proposed that gravity was not a force, as Newton and his followers believed, but a matter-induced distortion of space-time itself. Bodies moving under the influence of gravity, according to Einstein, are truly force-free and move in straight lines (called geodesics), that is lines as straight as a gravity-warped geometry will allow. Einstein's general theory requires not only that space be curved but time as well. Curved time means that how fast a clock ticks depends on where it's sitting.

Curved space-time is called "Riemannian geometry" after Georg Riemann, a nineteenth-century German mathematician who developed the symbolic tools Einstein needed for dealing with curvature in many dimensions. The laws of curved space-time are embodied in Einstein's equations, which describe in detail how massive bodies distort space-time in their vicinity, and how objects move in this distorted space-time.

Curved space-time gives rise to complicated effects that only Einstein's general theory can describe, but in flat space Einstein's special theory suffices. In particular, the speed-of-light limit—one of the consequences of special relativity—holds wherever space-time is flat. But no matter how twisted space-time gets, any region, when viewed on a small enough scale, looks flat. This means that even in general relativity the speed-of-light limit holds locally, at every point in space-time. The presence of local speed limits outlaws short FTL trips, but because space-time on a large scale is actually curved, long journeys at effective superluminal velocities may still be possible. Even when all local speed limits are obeyed, certain space-time configurations may allow global violations of the Einstein limit.

If the Earth were flat, and a 55-mph speed limit were imposed world-wide, it would represent the fastest anyone could legally travel from point to point. However, the fact that Earth is actually curved permits faster effective trips between two locations without violating the local speed limit. By traveling through a deep tunnel at the 55-mph limit, one can move between two points on Earth's surface considerably faster than the official speed limit. Likewise in the twisted geometry of general relativity, a local speed limit can sometimes be globally violated by exploiting similar shortcuts in curved space-time—the fabled "space warps" of science fiction.

FTL travel via warped space-time is the fictitious mechanism for spaceship propulsion in *Star Trek* and other sci-fi adventures. Warped space-time is also responsible for the alleged stargates purported to exist inside certain kinds of black holes. Whether such stargates are in fact possible could be settled by solving Einstein's equations for a variety of realistic situations. However these equations are extremely complicated and have only been solved for simple idealized cases.

Furthermore, there are no black holes that we can examine in the laboratory that would help us develop an experimental intuition for black-hole phenomena to test the various interpretations physicists have placed on the novel twists in space-time predicted by Einstein's equations. (The nearest astronomical black-hole candidate is Cygnus X-1, more than 10,000 light-years away.)

There is no doubt that certain solutions of Einstein's equations contain stargate and time-travel possibilities. But do these solutions represent anything that could happen in the real world? As candidates for space-warp-based communication let's consider seven special solutions of Einstein's equations for curved space-time:

1. Gödel's rotating universe;
2. Tipler's infinite rotating cylinder;
3. Schwarzschild (massive) black hole;

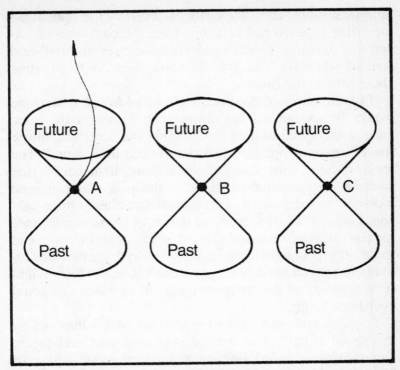

Figure 6-1. A family of light cones in flat space-time is shown. From location *A*, only events located inside the future light cone—events that are timelike-separated from event *A*—are accessible to travelers that obey the Einstein speed limit. To reach events *B* or *C*, separated from *A* by a spacelike interval, a person starting from event *A* would have to travel faster than light.

4. Reissner-Nordstrom (massive, charged) black hole;
5. Kerr (massive, rotating) black hole;
6. Kerr-Newman (massive, charged, rotating) black hole;
7. Super-extreme Kerr object (SEKO).

Gödel's Universe

In 1949, the celebrated Princeton mathematician Kurt Gödel discovered a solution of Einstein's equations that describes a Universe in which the tendency for collapse due to gravitational self-attraction is exactly balanced by the centrifugal force due to universal rotation. Like the conventional expanding universe, Gödel's universe possesses no privileged center—wherever you go, the stars seem to be rotating about you as the center.

If nature imposed no speed limit, all of future space-time would be accessible from the present. However the presence of the speed-of-light limit shrinks the accessible future space-time to a region called the "future light cone." The speed-of-light limit also defines a "past light cone" that contains all space-time regions in the past that can send messages to the present. Together, the past and future light cones contain all of space-time that is of physical relevance to this present moment. For this moment in time and space, the past light cone contains everything that has already happened about which you can know; the future light cone contains all the space-time regions to which you could ever hope to go.

The light-cone structure of space-time that is imposed by the speed-of-light limit acts to segregate past and future events strictly, confining each kind of event—past or future—to its own light cone Figure 6-1 illustrates the light-cone structure of flat space. One way to describe time travel—a breakdown of the strict segregation of past and future—would be to say that space-time becomes so contorted that the future light cone intersects the past light cone.

Consider a circular path in Gödel's universe that lies in a plane perpendicular to the rotation axis. Along this path the universal rotation of matter twists space-time in such a way that all light cones tilt in the direction of rotation. For circular trips whose radii exceed a certain critical value R, the light cones tip so much that the future light cone of one location intersects the past light cone of an adjacent location. For all circles with radii larger than R, it is possible for a traveler to move into the future, go around the loop, and enter his own past light cone, returning not only to the same location in space but also to the same moment in time

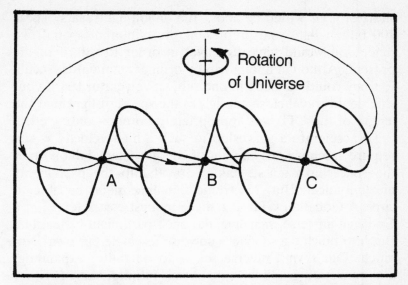

Figure 6-2 A family of light cones in a rotating (Gödel) universe is shown. Because light cones are tilted by a rotation-induced twist in space-time, traveler A can take a trip into the future, travel around a circular path, and end up in his own past without ever going faster than light. This kind of round trip, whose trajectory winds back into the past without ever getting spacelike (FTL), is an example of a closed timelike loop (CTL).

when he left. A rocket that moved along such a reentrant space-time path—called a "closed timelike loop," or CTL—would function as a time machine.

In Gödel's universe, the rotation rate must be fast enough to oppose the gravitational attraction of the universe's matter. Heavy universes have to spin faster. In order for centrifugal force to balance the matter density of our own Universe (including estimated dark matter), Gödel's model must make one complete turn every 70 billion years—a time comparable to our Universe's apparent age of 10–20 billion years. Our Universe must turn around at least this fast—call it the "Gödel rate of rotation"—to function as a natural time machine.

If it were spinning at the Gödel rate, our Universe's critical radius would be about 16 billion light-years and the shortest CTL would be about 100 billion light-years. These 100 billion light-years represent the minimum spatial distance you would have to travel in order to return to the present. Although a long trip by ordinary standards, such a journey could be made in one subjective year or less if your ship could travel close enough to the speed of light that the length of the CTL was appropriately Lorentz-contracted.

To accelerate a spaceship to such a high velocity would require an enormous amount of energy—many billion times the spaceship's mass, making travel into the past highly uneconomical. But, electronic signaling might be able to circle a Gödelian CTL at a more modest cost. Hence, in a Gödelian universe, signaling into one's past might be feasible. But how much does Gödel's universe resemble our own? Not much. Our own Universe seems to be finite, expanding, and nonrotating. Gödel's universe is infinite, static (neither expanding nor contracting), and rotating. Even if our Universe were rotating at the Gödelian rate, we could still not travel backward in time because our Universe is so small there isn't enough room to make the minimum 100 billion light-year CTL circuit. Gödel's infinite universe, on the other hand, permits round trips of any conceivable radius. For a universe as small as our own to be able to function as a natural time machine, it would have to rotate at least ten times faster than the minimum Gödel rate.

The feature of Gödel's universe most essential for time travel is rotation. The universe's rotation is what tips the light cones so that a traveler is able to enter his own past. In 1982, Paul Birch at Jodrell Bank, England, reported that surveys of distant radio sources showed systematic twists in the polarization of their light. Birch calculated that a universal rotation of about one-thousandth of the Gödel rate would suffice to explain his data. On the other hand, a universal rotation this large should produce a measurable anisotropy in the relic radiation from the Big Bang; light from the Big Bang should be more intense in certain directions. Cambridge University astrophysicists Collins and Hawking have argued that the observed high degree of isotropy of this radiation implies that our own Universe's rotation, if any, must be less than one-trillionth of the Gödel rotation rate.

Gödel's model shows without a doubt that general relativity predicts the possibility of time travel. If a universe rotates fast enough, its inhabitants could visit or signal their past selves by traveling around large circular orbits. However, it appears quite likely that our own Universe happens not to be constructed along Gödelian lines.

Tipler's Infinite Rotating Cylinder

In 1974, Frank Tipler, a mathematical physicist at Tulane University in Louisiana, showed that it is not necessary to rotate the entire universe to achieve time travel. An infinitely long massive cylinder spinning in such a way that centrifugal forces balance gravitational attraction can also tilt light cones outside the cylinder in a manner similar to the light-cone tiltings of Gödel's universe. When the surface of the cylinder travels at half the speed of light or greater, certain paths that encircle the cylinder become closed timelike loops. Although Einstein's equations have not been solved for finite cylinders, comparison with Newtonian solutions for the same geometry suggest that a region close to a finite rotating cylinder will also contain CTLs. Hence, Tipler argues, we might be able to create a time machine by spinning a massive skyscraperlike object at near-light speed.

In his book *Spacewarps*, John Gribbin calculates that 100-km-long, 10-km-radius cylinder with a mass equal to that of the Sun and rotating twice each millisecond, would meet Tipler's criteria. Gribbin points out how close this rotation rate is to some of the fast pulsars, implying that certain celestial objects that function as Tipler time machines might already exist.

For those who dream of building a time machine by spinning up a slender asteroid, Tipler warns that a finite rotating cylinder is vulnerable to axial collapse. Although centrifugal force prevents it from collapsing radially, the cylinder's own rigidity is all that resists the gravitational attraction along the rotational axis. For bodies heavy enough to act as time machines, no known form of matter can hold out against the massive cylinder's strong axial self-attraction. Under the influence of enormous gravitational forces the cylinder would turn into a spinning pancake. Tipler has shown on quite general grounds that any attempt to build such a time machine must end in some part of the machine collapsing into a singularity—a region of infinite gravitational curvature. In other words, before you can get your time machine to work, its own gravitational attraction always turns it into some sort of black hole. This dire fate need not be fatal to the time-travel enterprise, for are not black holes already renowned (at least in science fiction) for their ability to function as time tunnels?

Schwarzschild Black Hole

In 1917 Einstein & GROSSMAN formulated his THEIR general relativity equations, which describe how the space-time continuum is warped by the presence of mass. Before the year was out, German astronomer Karl Schwarzschild published a solution describing the Einsteinian distortion of space-time around a spherically symmetric, nonrotating mass. Schwarzschild's solution possesses the peculiar feature that for distances less than a certain radius R ($R = 2GM$), gravity does not allow light to escape. If the mass responsible for the distortion lies entirely within this "Schwarzschild radius," it is totally invisible, a black hole in space. For a solar-mass object, the Schwarzschild radius is about 3 km, meaning that if the Sun (whose present radius is about 700,000 km) were compresse either by its own gravity or by some external force to a sphere the size of San Francisco, it would suddenly turn black, all its light trapped forever inside its Schwarzschild radius. A black hole as massive as Earth would be only about 2 cm in diameter—the size of a large grape.

To display the time-travel possibilities of black holes, physicists use a graphic tool called the Penrose diagram.

Like the Mercator projection that makes Greenland loom larger than America on a flat map, the Penrose diagram distorts some features of space-time in order to picture other features correctly. The Penrose diagram grossly distorts sizes and distances but accurately portrays the space-time signatures (whether connections between space-time events are timelike, lightlike, or spacelike) of paths drawn on the map. All timelike paths, for instance, make an angle between 0 degrees and 45 degrees with the vertical; all lightlike paths are tilted at 45 degrees. All other paths are spacelike. The Penrose diagram shows at a glance whether a certain path is allowed (timelike or lightlike) or forbidden (spacelike—accessible only by FTL process) by the causal ordering postulate of special relativity.

Let's consider first a Penrose diagram for a body like the Sun, which has not yet condensed into a black hole. This diagram is two-dimensional, the vertical axis standing for time and the horizontal axis for distance r away from the Sun's center. In order to represent spatiotemporal regions at infinity on a finite diagram, these regions are compressed and distorted so all of space and time fit inside a simple triangle. The boundaries of this triangle correspond to five classes of infinity. Fig. 6-3 is a Penrose diagram illustrating these five infinities in flat space relative to an ordinary object such as the Sun.

Consider the timelike past infinity P, the ultimate source of all timelike trajectories in the world. Every material object, everything that travels slower than light, when traced back to its origin comes out of this timelike past. On the Penrose diagram this vast region of pastness is represented by a single point P.

The timelike future infinity F, the ultimate destiny of all material particles—in the absence of black holes—is likewise represented by a single point F. The entire space-time story of our reference object—the Sun in this case—is mapped by a straight line running from P to F. The space-time story of all other objects—Earth, for example—is mapped by a curved line between F and P. Earth's line may

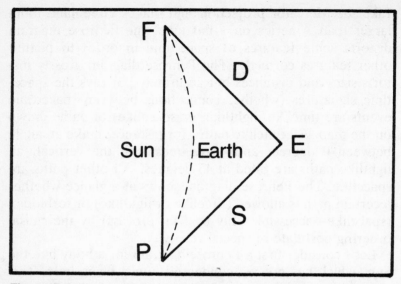

Figure 6-3. Penrose diagram centered on the Sun depicts the five classes of infinity relevant to a universe that contains no black holes. The edges of this diagram represent the timelike past and future infinities *P* and *F*, the lightlike past and future infinities *S* and *D*, and the spacelike infinity *E*.

twist about in the Penrose diagram but like all timelike lines, it must always make an angle of less than 45 degrees with the vertical (time) axis.

The lightlike past infinity *S* is the source of all light that is shining toward the Sun. These light sources are represented by a line *S* inclined at 45 degrees to the Sun's vertical timeline. All light rays originating at *S* are represented by lines tilted by –45 degrees and directed toward the Sun's timeline. Similarly, the lightlike future infinity *D* is the destination of all light that shines away from the Sun. This light's ultimate fate is represented by a line *D* inclined at –45 degrees to the Sun's timeline. All light traveling from the Sun to *D* is represented by lines tilted by 45 degrees running from the Sun's timeline to the *D* line.

To complete the cast of infinities, there's the spacelike infinity *E*, sometimes called the "absolute elsewhere." All very distant space-time regions inaccessible from the Sun by subliminal signals are represented by a single point E at the intersection of the past and future lightlike infinities S and D.

In the presence of a black hole, the ultimate destiny of light and material objects is expanded in scope. Instead of opting for one of the five infinities, an object can plunge into a black hole and effectively leave our universe. At the black hole's center—at least in the Schwarzschild case—lies the dreaded singularity, a region of infinitely stressed space-time where all matter is crushed out of existence, where space-time is so intensely bent that physics as we know it comes to an end. On the Penrose diagram the singularity is customarily represented by a jagged line—a row of shark's teeth to warn the space traveler of imminent danger and to instill dread in the hearts of theoretical physicists.

A second peculiar feature of the black hole is the one-way membrane located at the Schwarzschild radius. Be-

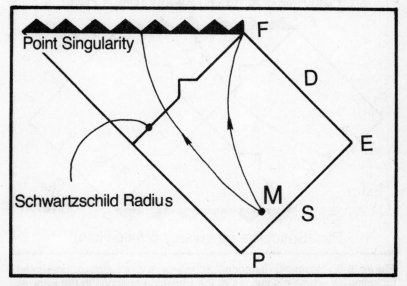

Figure 6-4. To the features common to a simple universe, the Schwarzschild black hole adds a spacelike future singularity where space-time curvature becomes infinite, plus a one-way membrane (enter only; no exit) located one Schwarzschild radius away from the singularity.

cause of strongly curved space and time near such a hole, objects can cross this membrane in one direction (going into the hole) but not the other.

Figure 6-4 shows the Penrose diagram for a Schwarzschild black hole. Max, the intrepid space pirate, looks at this Penrose map and sees that he has two choices. He can avoid the black hole and end up at F, the timelike future infinity, or he can cross the Schwarzschild radius and be sucked into the singularity. Max can travel into the distant future, or into the black hole. No FTL travel options here.

Figure 6-4 does not, however, picture the complete range of spherically symmetric solutions to the Einstein equations. The full solution containing all possible black hole options and connections is shown in Figure 6-5. Two features of Figure 6-5 are especially noteworthy: (1) the "white hole," and (2) "universe #2."

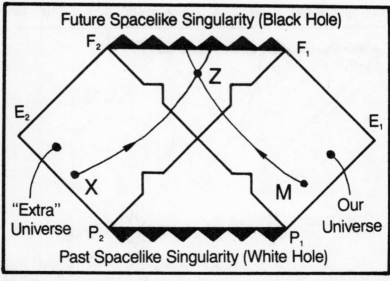

Figure 6-5. Maximally extended Schwarzschild solution is presented. Including all possible solutions of the Einstein equations in the Penrose diagram for a spherical mass distribution gives rise to an "extra" universe equipped with its own one-way membrane, plus a past spacelike singularity, the so-called "white hole," surrounded by an exit-only/no-entrance one-way membrane.

At the bottom of this complete Penrose diagram is a second singularity, a time-reversed copy of the singularity at the center of the simple black hole. Instead of sucking in matter, the time-reversed singularity spews out matter. Like its attractive brother, this repulsive singularity is surrounded by a Schwarzschild radius—a one-way membrane in space-time—but this membrane keeps incoming matter out rather than trapping it inside. Singularities of this repulsive kind—called "white holes"—are necessary features of the full solution to Einstein's gravity equations for the same reason that advanced and retarded solutions exist for Maxwell's electromagnetic equations. Because Einstein's equations are time-symmetric, the presence of black holes in the formalism automatically entails the possibility of white hole solutions.

Just because white holes exist as solutions to certain equations does not mean that they exist in our Universe. Some physicists have attempted to identify white holes with quasars—tiny superbright sources of light kindled near the era of galaxy formation—or with active radio sources, which have been observed to spew out huge energetic jets of matter. But most physicists believe that these phenomena can be explained as the results of more conventional processes. Hence there may be nothing in the visible Universe that corresponds to these white hole solutions.

There are no white holes, with one conspicuous exception. The unique event that gave birth to our entire Universe—the so-called Big Bang—possesses many of the same bizarre features as a time-reversed singularity, spewing out an ever-expanding sphere of matter from a space-time point of infinite gravitational curvature. If our Universe contains enough matter to bring about a future collapse—the ultimate Big Crunch—then the Universe's Penrose diagram would resemble Figure 6-5. The story of the Universe would then consist of a long journey from a big repulsive singularity to a big attractive singularity. In the full solution to the Einstein equations, along with the white hole, a "second universe" appears alongside our own, containing its own white and black hole pair. What is the nature of this new universe?

Universe #2 was discovered in 1935 by Albert Einstein and Nathan Rosen when they attempted to use the Schwarzschild solution as a model for elementary particles. Figure 6-5 shows that universe #2 and our own Universe are mutually inaccessible except via signals that travel faster than light. The universes are indeed connected—for instance, by the horizontal line that bisects the Penrose diagram—but all connections that can be made between the two universes are spacelike. The superluminal spacelike link between two otherwise inaccessible worlds is called an "Einstein-Rosen bridge" or "wormhole."

In the absence of experimental examples of black hole activity, theoretical physicists just have to guess how to interpret universe #2. Einstein and Rosen supposed that it might really be a distant part of our own Universe and envisioned the throats of the two wormholes to be the positions of a pair of elementary particles each with opposite electric charge. Although this Einstein-Rosen model is occasionally revived, the notion that elementary particles can be constructed out of twists of space-time has not been successful in explaining the behavior of matter in the small. Physicists generally treat the Einstein-Rosen proposal as a mere curiosity.

Is universe #2 a separate reality or merely a distant region of our own Universe? In the Big Bang model of creation and collapse, what corresponds to universe #2 is all matter lying behind the event horizon, matter we cannot communicate with because at the time of the Big Bang, this matter shot off in the opposite direction from us at the speed of light. This picture of our universe as a white hole/black pair certainly suggests that universe #2 is merely a distant region of our own. But for small black holes inside our universe, this identification might not be valid: The Einstein-Rosen bridge could conceivably connect to a universe entirely outside of our own Big Bang scenario. In the black hole case, the inhabitants of universe #1 cannot communicate directly (except via forbidden FTL channels) with the inhabitants of universe #2. However these folks do have an opportunity to receive messages or envoys from one another after they enter the black hole. Behind the Schwarzschild radius Max can meet his alien counterpart Xam (say at space-time event Z) and engage in a brief cultural exchange before they both eventually plunge into the deadly singularity. Although information can be exchanged between occupants of separate universes during occasions of mutual suicide, nowhere is any superluminal communication possible. Despite its bizarre features—white holes, extra universes, and spacelike wormholes—the maximally extended Schwarzschild solution offers no opportunities at all for FTL signaling or time travel.

Reissner-Nordstrom Black Hole

Nature links up her scattered particles via only two long-range forces, gravity and electromagnetism. Gravity is always attractive but electric charges (of like sign) exert repulsive forces on one another. At about the same time as Schwarzschild devised his spherical solution, German physicist Heinrich Reissner and Finnish physicist Gunnar Nordstrom solved Einstein's equations for a body endowed both with mass M and electric charge Q. For sufficiently large mass densities the Reissner-Nordstrom solution describes a spherically symmetric black hole whose Penrose diagram appears as Figure 6-6. In addition to white holes and second universes familiar from the Schwarzschild solution, two strikingly new features appear in the R-N black hole diagram: (1) timelike singularities and (2) "paper-doll topology."

Another important difference between these two types of black holes is that in the R-N solution two one-way membranes rather than one separate ordinary space from the central singularity. Figure 6-6 shows Max leaving location M, choosing route C, passing through both one-way mem-

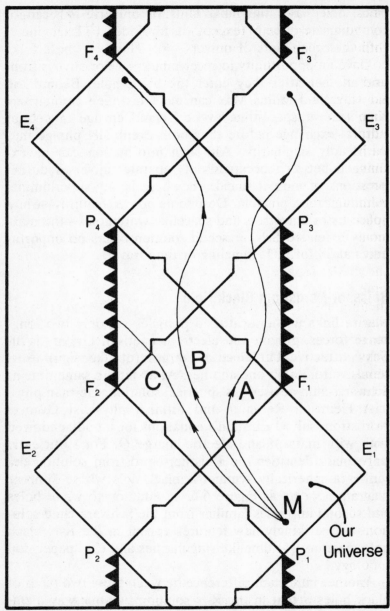

Figure 6-6. The Reissner-Nordstrom solution for a black hole with charge Q and mass M. A new feature of the R-N solution is its "paper-doll topology," an infinite proliferation of "extra" universes stitched together by timelike (avoidable) singularities.

branes, and crashing into the Reissner-Nordstrom singularity, which, like the Schwarzschild singularity, is a region of infinite space-time curvature.

Unlike the Schwarzschild singularity, which is spacelike and stretches across all future paths like a deadly net that no traveler can hope to escape, the R-N singularity is timelike, parallel to time's flow, and can be avoided by adroit space-time navigation. For example, if Max chooses path A, he avoided the R-N singularity and ends up safe in universe #3. This new kind of singularity is due to the presence of two one-way membranes in the R-N solution. Although time and space are both dimensions of Minkowski space-time, time is distinguished from the spatial dimensions by its unique metric signature. One important property of a one-way membrane is that crossing it interchanges the metric signature of the distance coordinate, r, and the time coordinate, t. Thus whenever Max traverses a one-way membrane, his time changes into space, and vice versa. In black hole physics, an odd number of such reversals leads to a spacelike (unavoidable) singularity like that in the Schwarzschild solution; an even number of membrane reversals leads to a timelike (avoidable) singularity, as in the Reissner-Nordstrom case.

Another big difference between these two species of black holes is the appearance of an infinite number of extra universes in the R-N case compared to only one extra universe in the Schwarzschild solution. These extra universes (each with its own singularity) repeat forever in both past and future time directions like a strip of paper dolls cut out of folded paper. Furthermore, these universes are all accessible: A space traveler following a COP-allowed path—a path that is always timelike—can avoid all singularities and end up in universe #3 (Route A), universe #4 (Route B), or in some other alternate reality of his choice.

For the traveler's convenience, each R-N singularity acts both as black hole and white hole at once (a kind of space-time trampoline?), that is, each singularity possesses both entrance and exit one-way membranes. Although this symmetric arrangement of unidirectional membranes allows one to escape from an R-N singularity, it does not permit direct return to the universe of origin unless one of the extra universes happens to be a different view of our home Universe. If these extra universes are really parts of our own, as viewed from R-N white holes scattered throughout space, then such a black hole (or holes) could function as a stargate for FTL travel and/or for trips into the past.

Trips into the past give rise to closed timelike loops (CTL) or what Cambridge physicist Brandon Carter calls "vicious circles." Carter calls any set of space-time points from which a CTL (or vicious circle) can be launched, a "vicious set." He distinguishes two kinds of vicious sets: the "trivially vicious" sets that arise only if you identify the extra universes as distant parts of our own Universe, and the "flagrantly vicious" sets in which CTLs persist no matter what the interpretation of the extra universes. In Carter's terminology, all space-time points within traveling range of a Reissner-Nordstrom black hole form a trivially vicious set. The R-N solution has no flagrantly vicious points— CTLs that can't be wished away. The entire Gödel universe, on the other hand, is an example of a flagrantly vicious set since every point in Gödel's universe can be the starting point of a CTL.

To construct an R-N black hole you need a lot of matter highly charged with a single electric polarity (either very positive or very negative). However the intense electric fields created by these isolated charges would certainly ionize any matter present into positive and negative charges which would move at once to neutralize the excess charge on the black hole. Thus in a universe like our own, made of matter in which both signs of charge exist in roughly equal amounts, formation of a Reissner-Nordstrom black hole is highly unlikely. Hence the type of stargate that the R-N solution contains is almost certainly irrelevant in our kind of universe.

In the Schwarzschild solution, the link between universes is in a sense slammed shut by the universally attractive nature of gravity. The opportunities for exotic travel afforded by the Reissner-Nordstrom solution result from the repulsive nature of the electrostatic force which, when it opposes the gravitational contraction, "holds open the door" to other universes.

Like the electrostatic force, the centrifugal force associated with a rotating body acts to oppose gravitational collapse. But unlike electric charge, which is generally ineffective in the evolution of celestial bodies, rotation is a common formative feature of stars, planets, and galaxies. Hence the solution of Einstein's equations for a massive rotating body is especially relevant to the question of whether black holes can really act as stargates to other universes.

Kerr Black Hole

After the Schwarzschild and Reissner-Nordstrom solutions were published, almost half a century elapsed before Roy Kerr, a physicist from New Zealand, solved Einstein's equations for a rapidly rotating mass. The long delay in obtaining such an important result is a measure of the complexity of Einstein's equations, and suggests that they contain even more surprises.

In addition to the attributes already present in the R-N solution, the Kerr black hole contains three new features: (1) the ring singularity (2) the "negative" space-time extension and (3) the flagrantly vicious set. Like the R-N solution, the Kerr black hole possesses a paper-doll topology and timelike (avoidable) singularities. Unlike the other black holes, the Kerr hole's singularity is not a point but takes the form of a ring encircling the axis of rotation. Because of the singularity's ring shape, it is possible to travel to the exact center of the Kerr hole without encountering a region of infinite curvature. This central "softness" of the ring singularity is symbolized in Figure 6-7 by rounding off the singularity's shark teeth in the Kerr black hole's Penrose diagram.

By allowing access to the black hole's center, the ring singularity gives rise to another peculiar effect, the "nega-

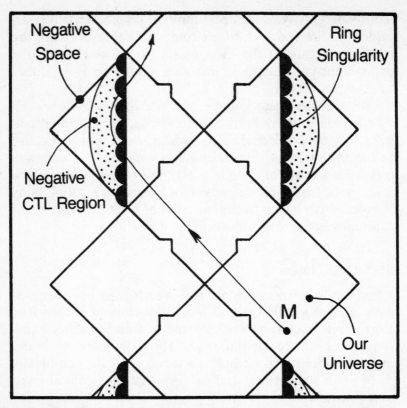

Figure 6-7. The Kerr solution for a black hole with angular momentum *A* and mass *M*. A new feature of the Kerr solution is the "negative space"—a repulsive-gravity universe lying on the other side of the ring singularity. Inside this negative space exist regions from which you can launch trips into the past (negative CTL regions).

tive" space-time region. If we take the Einstein equations seriously, we find that as a traveler passes through the hole's center ($r = 0$), the product $G \times r$ (where G is the universal gravitational constant) suddenly becomes negative. This means either that r is negative—that somehow a meaning should be attached to being a negative 5 km away from the center of a sphere—or G is negative, and we have entered a region of negative (that is, repulsive) gravity.

In the absence of any experimental data about the insides

of real Kerr black holes, we are free to guess the significance of the negative space-time that is lurking on the "other side" of the ring singularity. Whatever interpretation we give to this negative space-time, the Kerr solution is not complete until we assign this space its own set of fivefold infinities, as in Figure 6-7. The presence of this negative space gives rise to a third peculiar feature of the Kerr solution—the existence of locations from which it is possible to travel backward in time.

While orbiting the axis of rotation inside the ring singularity, no causal anomalies occur as long as one does not try to go through the ring. However upon crossing the plane of the ring, one enters the negative space, and travels backward in time to an extent that depends on the number of times one circles the axis. Thus Max can travel around a closed timelike loop by dipping into negative space, and actually return from a trip before he leaves. The region of negative space near the ring singularity qualifies, in Brandon Carter's terms, as a vicious set of the flagrant kind.

However, despite the presence of a CTL-infested region near the ring singularity, Max cannot simply enter a rotating black hole, go back in time, and return to ordinary space before he left. The pair of one-way membranes he had to pass through to reach the singularity prohibits any such direct backtracking. Because of these one-way membranes, the only way to escape a Kerr black hole is to exit into a different universe from the one you left. This exit universe may or not not be the same as the entrance universe. Anyone who plans to use a Kerr black hole for time travel—as did Siborg, the alien "Cornell professor"—should get hold of a good set of roadmaps to the extra universes before he embarks.

Kerr-Newman Black Hole

A few years after Roy Kerr published his rotating black hole solution, Ezra Newman and his colleagues at Pittsburgh extended the Kerr solution to the case where the rotating mass is also charged. The Penrose diagram of a

Kerr-Newman black hole (Fig. 6-8) closely resembles that of a plain Kerr hole with one addition: The region inhabited by flagrant CTLs now extends through the ring singularity to contaminate ordinary positive space. Hence the Kerr-Newman black hole offers the opportunity for time travel without the necessity of diving through the ring singularity into the problematic region of negative space-time; time travel is possible without leaving ordinary space-time— that is, if you consider the space-time inside a highly charged, rapidly rotating, extremely massive object "ordinary."

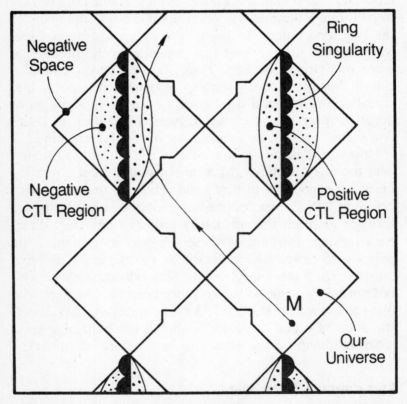

Figure 6-8. The Kerr-Newman solution for a black hole with angular momentum *A*, charge *Q*, and mass *M*. Besides all the features of a Kerr hole, the K-N solution possesses time-travel options (flagrant CTLs) in ordinary positive space as well as in the negative-gravity spaces behind the ring singularity.

Super-Extreme Kerr Object (SEKO)

A Kerr black hole is completely characterized by the values of just two parameters, the hole's mass, M, and its angular momentum, A. Although they may have had wildly different histories, two black holes with the same value of M and A are observationally indistinguishable. Because it is completely defined by just a few parameters, a rotating black hole is actually a very simple object, a sort of gigantic elementary particle.

In the Kerr black hole the gravitational attraction due to the mass M is opposed by the centrifugal force associated with the angular momentum A. The gravitational attraction dominates as long as M is greater than A. As A increases the two membranes move closer together and when $A = M$ the membranes fuse. This situation in which the black hole's mass is precisely balanced by the hole's angular momentum is called an "extreme" Kerr hole.

If the angular momentum increases further so that it exceeds the object's mass, the one-way membrane vanishes and the ring singularity is exposed. The absence of a membrane means that a space traveler can come in from outside, visit the singularity, and then return to Earth. A strong prejudice exists among physicists against such so-called "naked singularities" because a singularity is a place where the laws of physics fail. As long as such outlaw regions are safely tucked away behind one-way membranes, their blatant lawlessness cannot escape to contaminate our Universe. Such imprisoned singularities are effectively shut up in another universe, so we can still regard the laws of physics as universally valid. To preserve the rule of law in our Universe, physicists have proposed a cosmic censorship principle that simply forbids the presence of naked singularities in the real world despite their appearance in certain solutions of the Einstein equations. Great effort has gone into making the cosmic censorship principle plausible but a proof that every singularity must be surrounded by a one-way membrane has so far eluded us.

*   *   *

A collapsed system whose angular momentum exceeds its mass might be called a super-extreme Kerr object (SEKO). A SEKO possesses a naked ring singularity, one negative gravity universe, and turns all of space-time into a flagrantly vicious set.

Since the SEKO's singularity is of the ring variety, and timelike as well, a space traveler can pass through the singularity without encountering an inevitably fatal region of infinite space-time curvature, and enter another universe where gravity is repulsive. Only one such universe is present here; a SEKO does not possess a paper-doll topology. The SEKO's most interesting feature is that by circumnavigating the singularity in the proper direction, one can go back in time. Moreover, all regions of space-time become accessible via such journeys, not just regions inside black

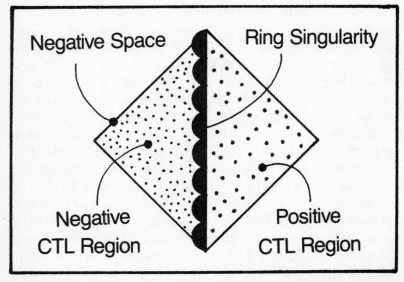

Figure 6-9. When the angular momentum *A* of a Kerr hole exceeds its mass *M*, the hole's one-way membranes disappear, exposing its naked ring singularity. This super-extreme Kerr object (SEKO) turns all of space-time into a flagrantly vicious set. If our Universe contains just one SEKO, any place at all could be the starting point for a trip into the past.

hole membranes. In principle one can approach a SEKO, fly around it a few times, and end up anywhere in the past—no need to travel through the singularity or travel into new universes of dubious authenticity. Since these trips into the past can be accomplished from anyplace at all—or at least anyplace from which you can reach the SEKO in a reasonable time—all of space-time becomes, in Carter's terminology, a flagrantly vicious set. If space-time is really flagrantly vicious, then from where you are sitting right now, a possible passage exists—could you but find it—to any event in the past.

Tipler's rotating cylinder would also produce a similar kind of flagrantly vicious space-time, but such a cylinder is unstable to axial collapse. A SEKO, on the other hand, is stable, as far as we know, a more robust version of Tipler's rapidly rotating time machine.

Real-Life Relativity

These examples show without a doubt that general relativity permits faster-than-light travel, rapid transit between distant universes (stargates), and trips backward in time. These solutions, however, are highly idealized and bear only a tenuous relation to what might actually go on in the world. For instance, Gödel's solution shows that time travel is possible if the universe is rotating fast enough. But our Universe doesn't seem to be rotating at all.

Tipler's rotating cylinder is unstable against axial collapse. No matter what sort of material you make it out of, as you try to assemble the cylinder, it collapses under its own weight into a spinning pancake, or into a black hole. A spinning cylinder would make a good time machine but because matter is not rigid enough there seems to be no way to construct a cylinder of the Tipler variety.

The black hole solutions considered here are of the so-called "exterior, source-free" type. These solutions represent the space-time outside the matter distribution that creates the hole. The matter itself does not appear in the solution because it is presumed to have collapsed into the singularity. Furthermore, all maximally extended solutions invoke the concept of a "white hole." In these solutions the history of a black hole begins with its birth out of a white hole and (except in the Schwarzschild case) continues with eternal oscillations between white and black hole incarnations, spewing out a repetitious space-time pattern that I call a "paper-doll topology."

In real life, black holes are not expected to grow out of white holes but result from accretion of ordinary matter. Most physicists believe that the presence of this infalling matter chokes off the central portion of the idealized hole, where all the interesting time tunnels meet, resulting in a rather prosaic black hole history, like that illustrated in Figure 6-10, rather than the more flamboyant holes previously discussed.

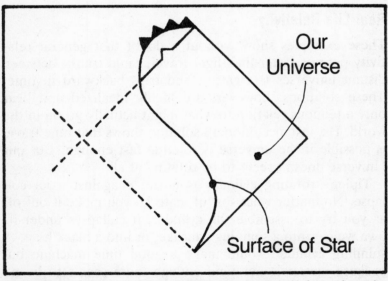

Figure 6-10. Realistic black hole? When a black hole forms from the gravitational collapse of a star, the region to the left of the Penrose diagram where all the interesting time-travel possibilities might be located, is filled up with the matter of the infalling star thus invalidating the exotic solutions previously considered, which assumed a space entirely free of matter.

A simple argument put forth by Douglas M. Eardley, a physicist at CalTech, probably eliminates white holes as objects that could really exist for any length of time in our Universe. Light traveling away from the Schwarzschild radius of a white hole will be red-shifted (decreased in energy) to zero energy by the hole's powerful gravitational field. Similarly, light traveling toward the white hole is blue-shifted (increased in energy) to an infinite extent, an effect Eardley terms the "creation of a blue sheet." The gravitational pull of this infinitely heavy light will collapse it into a black hole. Formation of a blue sheet effectively smothers any incipient white hole unlucky enough to attempt to expand into our Universe. Eardley's argument does not prevent white holes from forming in empty universes—universes devoid of light. But universes like our own contain lethal amounts of light and matter which will form fatal blue sheets that smother infant white holes in their cradles. Fortunately for us, Eardley's argument also does not exclude the production of our Universe itself as a white hole, since this event took place in an arena not only empty of light and matter, but devoid of space and time as well.

Another feature of idealized black holes essential for their role as stargates is the presence of timelike (avoidable) singularities. While these singularities are certainly predicted by general relativity, when quantum effects are taken into account, such singularities are probably obliterated. In the vicinity of such singularities the energy inherent in strongly warped space-time will be converted into an infinite number of quantum particles which proceed to travel into the future where they gravitationally collapse into a spacelike singularity. This quantum-induced conversion of singularities from timelike to spacelike acts like a trap door to close off the tunnels to other universes. Timelike singularities are permitted by general relativity but are unstable as far as quantum theory is concerned; quantum effects turn them into singularities of the spacelike variety which then remain stable features of space-time.

Since one of the essential features of a SEKO is its timelike singularity, quantum effects may quash the use of such objects as stargates. Also SEKOs may be difficult or impossible to produce by spinning up ordinary black holes. One can imagine two ways of increasing the angular momentum A of a black hole with the intent of making it exceed its mass M. You could either shoot in mass along a glancing trajectory or shoot in spinning objects that possess a high ratio of A to M. The electron's angular momentum, for instance, is almost a trillion times its mass, and other elementary particles possess similarly high A to M ratios.

However, calculations on the forces and permissible orbits near Kerr black holes show that the number of glancing orbits by which the $A{:}M$ ratio can be increased dwindles to zero as A approaches M. This means that as the hole's angular momentum increases it becomes difficult and then impossible to find an orbit that puts in more spin than mass. Furthermore, the shooting in of rapidly rotating objects is inhibited by a strong repulsive force that prevents entry by pushing such objects away from the hole if their spin is oriented in such a way as to increase A.

Although these arguments seem to preclude the use of idealized black holes and SEKOs as time tunnels, they do not eliminate the possibility that certain other configurations of matter might be discovered which, under realistic conditions, might embody some of the exotic space-time travel options permitted by Einstein's equations. The search for realistic solutions of the Einstein equations that contain flagrant CTLs or accessible extra universes seems a worthy task for large fast computers. One might begin by investigating complex configurations of rotating black holes, the region, for example, between two corotating Kerr holes orbiting just outside of each other's one-way membrane.

Einstein's equations of general relativity do not exclude the possibility of FTL travel. In fact, for certain idealized situations, these equations actually necessitate FTL behavior. In light of this live loophole, the search for realistic solutions of the Einstein equations (including quantum effects) that contain stable FTL possibilities does not seem to be an entirely futile venture. Lurking in the complexity of these equations may be the key to easy journeys to other galaxies courtesy of space-time wrinkles of the superluminal kind.

CHAPTER 7

Deviant Denizens of the Timestream: Tachyons, Antiparticles, and Neutral Kaons

As they ran backwards, what was to men an expanding universe appeared to the tachyons as a contracting one. . . . Gordon's measurement of the tachyon flux, integrated back in time showed that the energy absorbed from the tachyons was enough to heat the compressed mass. This energy fueled the universal expansion. So, to the eyes of men, the universe exploded from a single point because of what would happen, not what had. Origin and destiny intertwined. The snake ate its tail.

—Greg Benford, *Timescape*, 1980

The idea of a particle that travels faster than light was first conceived by Munich professor Arnold Sommerfeld around the turn of the century. On the basis of Maxwell's electromagnetic theory, Sommerfeld argued that, in contrast to ordinary particles which speed up as they gain energy, these FTL particles would speed up as they lose energy. In the early sixties, Sho Tanaka (Japan), Olexa-Myron Bilaniuk (Ukraine-USA), George Sudarshan (India), Yakov Terletskii (USSR), Gerald Feinberg (USA), and others revived interest in superluminal particles they called "*B*-matter," "meta-matter," or, more commonly, "tachyons."

Einstein's theory of special relativity contains two barriers that prevent ordinary particles from going faster than light. The first barrier—the causal ordering postulate—forbids all spacelike causal connections. Since it covers more space than time during its journey the trajectory of an FTL particle is certainly spacelike, but is such a connection always causal? It certainly seems plausible that a particle traveling from A to B could act as causal influence or carry a message between these two locations. Unless some mechanism that dilutes the causality of these particles is invoked, the COP principle would surely forbid FTL particle motion.

The second barrier to FTL motion is the relativistic mass increase. As its speed increases, a particle becomes heavier, hence more resistant to further acceleration. At the speed of light, a particle's mass is infinite. Hence an infinite quantity of energy is required just to reach the speed of light; no amount of energy, however large, can make a particle move faster than light.

Of these two principles, the second is more serious. the COP principle is somewhat abstract. The concept of "cause and effect," for instance, is loosely defined and may contain hidden loopholes. Furthermore the COP principle itself might not be universally valid. Although no spacelike causal connections have yet been discovered, a single counterexample would suffice to abolish our prejudice that the causal direction of all processes must be the same for all observers.

On the other hand, for high-energy physicists, the relativistic mass increase is a familiar and well-tested phenomenon. In laboratories all over the world the mass increase of elementary particles as they approach the light barrier is a matter of daily experience.

However, quantum theory, the most recent model of how matter behaves, may provide loopholes through which the mass-increase light barrier might be circumvented. In addition to describing how particles move around, quantum theory describes the creation and annihilation of particles. When two automobiles collide, their kinetic energy is converted into heat and twisted metal. When two quantum particles collide, their kinetic energy can be transformed (courtesy of Einstein's $E = mc^2$) into new particles. Out of a quantum collision, any particle consistent with energy, momentum, charge conservation laws, and a few more exotic conservation laws can emerge. In fact, after a large number of collisions has taken place, any particle that is even remotely possible must be created, according to a kind of quantum "totalitarian principle" that CalTech physicist Murray Gell-Mann expressed this way: "[In a quantum collision] anything not forbidden is compulsory."

If we take this quantum totalitarian principle seriously, we can imagine bypassing the mass-increase barrier not by accelerating a particle to light speed and beyond, but by simply putting enough energy together in one place to create a particle that right from the start travels faster than light. If we could manage to create a superluminal particle in a quantum collision, what would it look like?

Tachyons

Special relativity describes the behavior of particles traveling at any velocity, v. If we let v be greater than c, these equations should also describe the trajectories of tachyons. According to this plausible extension of relativity, tachyons differ from their slower-than-light cousins (tardyons), and particles that travel merely at light speed (luxons) in three important ways:

1. A tachyon possesses an imaginary rest mass; that is, the square of its mass is a negative number.
2. As it loses energy, a tachyon speeds up—a result already anticipated by Sommerfeld in 1904. Once a tachyon has lost all its energy, it must travel at an infinite velocity; that is, it occupies every point along its trajectory at the same time. Olexa-Myron Bilaniuk and E. C. George Sudarshan, authors of an early treatise on tachyons, call a particle that dwells in this strange state of omnipresence—zero energy/infinite velocity—a "transcendent" tachyon.

3. To slow a tachyon down requires the addition of energy. To slow a tachyon down to light speed requires an infinite quantity of energy. Thus for a tachyon, the speed of light is a lower limit to its velocity. Once a tachyon, always a tachyon—such particles can never go slower than light.

Critics have dismissed tachyons as unphysical entities on account of their imaginary rest masses. Tachyon defenders counter that, since a tachyon can never be brought to rest, tachyon rest mass is not experimentally meaningful and does not need to be represented by a real number. A tachyon's energy and momentum, on the other hand, are measurable quantities, and are, represented accordingly, by real numbers, that is, numbers whose squares are positive.

Tachyon fact #3 states that the light barrier has two sides. Just as brute force cannot make a tachyon out of a tardyon, likewise no FTL particle can be forced to slow down to subluminal speed. A pleasing symmetry exists between these two types of matter—except for quantum creation events, tachyons and tardyons each stay on their own side of the speed-of-light fence.

An ordinary particle (tardyon) cannot travel faster than the speed of light in vacuum. However light in most transparent materials has a phase velocity considerably slower than its vacuum speed. Hence it is possible even for a mere tardyon to actually exceed the velocity of light in such materials. This kind of FTL behavior is not prohibited by special relativity because the speed of light in vacuum is Einstein's universal speed limit not the speed of light in glass or in plastic.

In 1934, Russian physicist Pavel Čerenkov discovered that whenever an electrically charged particle travels faster than light's phase velocity, the particle gives off light—now called Čerenkov radiation. A fast particle's Čerenkov radiation is analogous to a jet plane's sonic boom when it exceeds the speed of sound. Čerenkov radiation is a kind of "optic boom."

Since tachyons always travel faster than the speed of light in a vacuum, a charged tachyon is expected to emit Čerenkov radiation continuously. Charged tachyons could be recognized by that special glow. Since it could never shut off its Čerenkov radiation by slowing down (remember a tachyon speeds up as it loses energy), the fate of a charged tachyon is to lose all its energy quickly, in a flash of Čerenkov radiation and enter the zero energy/infinite velocity or "transcendent" state. Since this state has no energy to spare, a transcendent tachyon can emit no more light. If it gains some extra energy, however, by colliding with an ordinary particle or another tachyon, it quickly emits that energy as light and reenters the transcendent state. Electrically charged tachyons act rather like phosphors—the active material in a television screen—which are molecules that give off a burst of light when excited by fast particles. The presence of transcendent charged tachyons might be revealed by luminous tracks or flashes of light in seemingly empty space.

If tachyons are uncharged, they would not produce Čerenkov light and like neutral tardyons may not interact strongly with conventional detectors. The properties of weakly interacting particles such as the neutrino are often inferred indirectly by adding up the total visible energy and momentum in a reaction and attributing any missing energy and momentum to unobserved neutral particles. The presence of invisible faster-than-light particles could easily be recognized by this missing energy method, because the energy of all conventional particles (tardyons and luxons) always exceeds or equals its momentum. Conversely, a tachyon's energy is always less than its momentum.

The excessive momentum carried by an FTL particle constitutes a unique tachyon signature. The production of a tachyon in a bubble- or spark-chamber can be inferred if energy-momentum accounting leads to a big deficit in the energy column. In a dozen or so accelerators around the world, thousands of pictures of high-energy collisions are taken every day, each one a potential portrait of a tachyon.

No inexplicable streaks of light or energy-deficient reactions have ever been detected, either accidentally or in special tachyon searches carried out at Amherst College, Brookhaven National Laboratory, and several other sites. Despite this apparent absence of tachyonic events, one might always hope that they will show up at higher energies. Perhaps there exists a "tachyon threshold" our accelerators have yet to surmount.

A natural source of high-energy particles lies directly over our heads. Cosmic rays, intensely energetic particles of uncertain origin, bombard Earth daily and account for almost half of our exposure to radiation from natural sources. (Radioactive elements in the air, rocks, and our own bodies, make up the remainder.) Some cosmic ray energies exceed that of particles in conventional accelerators by a factor of a million or more.

When a cosmic ray strikes the atmosphere, the resulting nuclear reaction is too complicated and inaccessible to analyze via energy-momentum accounting, but it does offer us a more direct method for detecting the presence of tachyons. The primary cosmic ray travels effectively at the speed of light. Its initial reaction at the top of the atmosphere produces a myriad of particles—called secondary cosmic rays—that likewise travel at near light speed. These particles interact with the air to produce still other particles. The end result of a single cosmic-ray impact is a sudden burst of particles traveling at near-light speed. From its origin at the top of the atmosphere, this so-called "air shower" takes about 20 μsec to reach the surface of Earth. If a tachyon happened to be produced somewhere inside the air shower, it would travel faster than the shower itself and reach Earth before the burst of slow tardyons.

A conventional air-shower detector consists of an array of particle detectors spread out over a large area. The magnitude and direction of the primary cosmic ray is inferred from the difference in arrival times of the shower's front in each of these detectors. If all the particle detectors receive the burst at the same time, for instance, then the primary cosmic ray was directly overhead. One way to search for tachyons in cosmic-ray air showers is to examine detector records and look for anomalous events that precede the main shower front by 20 μsec or less.

In 1973, Roger Clay and Philip Crouch in Australia carried out such a search for faster-than-light precursors with positive results. Because other experimenters have not been able to repeat their success and experiments of different design have failed to detect any FTL premonitory events in the forefront of large air showers, the experimental verdict so far seems to weigh against the existence of tachyons. As Columbia physicist Gerald Feinberg has quipped, the only place you can find a tachyon today is in the dictionary.

Not only have tachyons never been observed, but their theoretical underpinnings are far from secure. For instance, one consequence of the quantum revolution is that to every particle there corresponds a wave—the quantum wave function—that represents the probability of observing the particle at a particular location in space. From the energy-momentum relations for ordinary particles, Erwin Schrödinger and Paul Dirac showed 50 years ago how to construct their quantum wave functions. Following these quantum rules, it is not difficult to write the tachyon wave function by simply replacing the real mass in the conventional wave equations with the imaginary mass characteristic of FTL particles.

But according to Louisiana State's Loris Robinet, one of the many physicists who have studied the solutions to the tachyon wave equation, the quantum wave for FTL particles always travels slower than light. For ordinary particles the velocity of the particle and the velocity of its quantum wave are the same. This is as it should be for the wave represents the probability of observing the particle. But for tachyons the situation seems to be different: tachyon as particle travels FTL; tachyon as wave respects the Einstein limit. The discordant velocities of tachyon particle and tachyon wave make it difficult to construct a consistent quantum theory of tachyons. Thus the very theory that opens a loophole for FTL particles (via the particularly quantum effect of particle creation) may have slammed the door shut on these same particles.

Although most physicists today place the probability of the existence of tachyons only slightly higher than the existence of unicorns, research into the properties of these hypothetical FTL particles has not been entirely fruitless. Much ingenuity has been exerted in the attempt to integrate FTL particles into polite society by dreaming up ways to eliminate the unsavory time-travel paradoxes that have prejudiced physicists from the start against particles of the superluminal persuasion. To disarm these prejudices, tachyon researchers have tried hard to show that although these particles did indeed travel faster than light, some special feature in nature will prevent their use as channels for backward-in-time signaling.

An early attempt to domesticate tachyons was Bilaniuk and Sudarshan's "reinterpretation postulate." They showed that whenever a tachyon world-line went backward in time, its energy also becomes negative. A negative-energy particle traveling backward in time can, however, always be reinterpreted as a positive-energy particle traveling forward in time. If such a reinterpretation of reality is allowed, then tachyons will always seem to travel forward in time carrying positive energy. However for the same tachyon traveling between locations A and B, one observer will say that A emitted the tachyon and B received it, while another observer will insist that it was B that sent the tachyon and A that received it. The reinterpretation principle achieves one-way time travel for tachyons by giving up the observer-independence of cause and effect. Bilaniuk and Sudarshan's reinterpretation principle violates the COP rule, however, because, for tachyon events, what is cause and what is effect depends on the observer.

The reinterpretation principle was criticized by Benford, Book, and Newcomb at Berkeley, who argued that the notion of cause and effect could not so easily be relativized.

If tachyons can be used for sending messages either by modulating a tachyon source or by selectively absorbing tachyons from a constant natural source, then the person at A who carries out this modulation is undeniably the cause, the person who decodes the message at B is always the effect. In other words, for any situation in which tachyons can be used for signaling, cause and effect can be determined independently of the observer's motion and the time ordering of A and B.

A simple thought experiment devised by Fritz Pirani, a mathematician at King's College, London, further weakened the reinterpretation principle. Pirani imagined three observers, each moving along one side of an equilateral triangle. He showed that if Tom sends a positive-energy tachyon (PET) to Dick, and Dick sends a PET to Harriet, then Harriet can send a PET back to Tom that will arrive before Tom sends his initial message. Pirani's example shows that it is possible to achieve a time-travel paradox using only positive-energy tachyons whose direction of travel cannot be reinterpreted. Although the reinterpretation principle may be able to eliminate temporal paradoxes for two FTL communicators, the "Pirani triangle" shows that this principle fails for certain three-tachyon telegraph systems.

Another proposed way to eliminate the paradoxical aspects of FTL particles is to postulate an absolute reference frame in which all tachyon motion "really" takes place. Only in this special frame can tachyons be emitted at all speeds and in all directions. Tachyon transmitters in other frames would experience peculiar constraints on their abilities to send FTL particles freely in all directions.

In the "primary reference frame" tachyons can go FTL but not backward in time. In this special frame all round trips from spatial location X, time T_1 back to location X, time T_2 must always end up in the future, that is, T_2 is later than T_1. The time ordering of any two events in space-time is the same for all observers provided the events are not spacelike-separated. Since this "later than" relationship between T_2 and T_1 is timelike, its order is preserved in all reference frames. The round trip will appear to end up in the future in all reference frames. The existence of a preferred tachyon frame eliminates trips into the past.

Some physicists, noting that all time-travel paradoxes arise from returning to a location before you left it, decided to eliminate the paradox by refusing to issue round-trip tickets to tachyons. If all tachyon sources can radiate only in a certain direction, along the so-called "tachyon corridor," tachyons would be allowed to go faster than light but could never produce temporal paradoxes because unidirectional tachyons cannot loop back to their source. However, as ingenious as these proposals may be, they all go against the central postulate of special relativity—that nature recognizes no special reference frames. If we want to be consistent relativists, then we should require that even for tachyons no moving frame or any spatial direction is better than any other.

Like pieces of a jigsaw puzzle, any new particles must have the right shape and size to fit into the laws of physics that we already know to be correct. Despite considerable effort by many scientists to tailor the properties of tachyons to suit the laws of relativity and quantum theory, the fit is still not good. No one has yet come up with a consistent model for a quantum particle that travels faster than light.

Antiparticles

Theory and experiment in physics often play a sort of leapfrog game. Sometimes experiment is out front producing new results unanticipated by the mathematicians; at other times, theory shoots ahead to predict an effect that the experimentalists scramble to verify or refute. One of the most impressive instances of theory-driven discovery was the prediction of a new kind of matter by Paul Dirac.

In 1926, Erwin Schrödinger devised his famous wave equation which governs the quantum probability waves associated with all material particles. Schrödinger's equation was immediately applied to many practical problems, including the structure of atoms, the scattering of elementary particles, and the nature of the solid state. For all its usefulness, however, the Schrödinger equation is "nonrelativistic," that is, because it is not formulated in "tensor language," it is valid only for particles whose velocities are slow compared to the velocity of light. Paul Dirac, working at Cambridge University, remedied this defect by constructing a relativistic wave equation for the electron—the first elementary particle to be discovered. His successful invention of the "Dirac equation" did more than provide physicists with a description of fast electrons. It led to many unexpected insights into the nature of the quantum world.

For example, the Schrödinger equation explains the spectrum of hydrogen quite well except that many spectral lines—single in the theory—actually appear doubled in the laboratory. Dirac's theory attributes this spectral doubling to the electron's intrinsic spin, which can point in only two directions (conventionally labeled "up" and "down"). Dirac might have achieved this result simply by tacking on another component to the ordinary Schrödinger description. Instead of taking this direct approach Dirac showed that the electron's two-fold spin is an inevitable consequence of the marriage of relativity and quantum theory. Dirac discovered that the electron's spin is, in a certain sense, a purely relativistic effect, like the retardation of a moving clock.

Dirac's double solution arises from the fact that in his theory the energy E of the electron is equal to the square root of a certain quantity X. But there are always two numbers whose square is X, a positive number $\sqrt{X}$ and a negative number $-\sqrt{X}$. Because of this doubling of energies, not present in the Schrödinger case, the Dirac equation predicts four different kinds of electron: spin-up and spin-down with positive energy, and spin-up and spin-down with negative energy. The properties of the positive-energy electrons correspond to ordinary electrons as seen in the laboratory, but where would one go to see a negative-

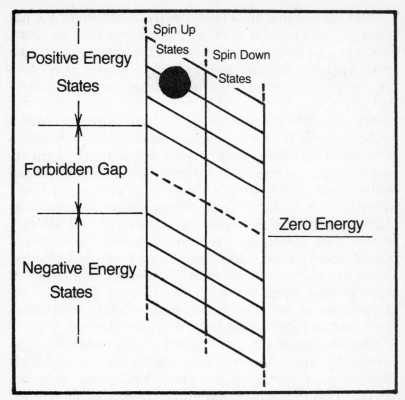

Figure 7-1. The Dirac equation for the electron has four types of solution: spin-up and spin-down electrons that can possess either positive or negative energy. The energy level diagram shown here is occupied by one positive-energy, spin-up electron. Dirac's unmodified theory predicts that such an electron is highly unstable: In a few billionths of a second it will drop into one of the negative-energy states while emitting a brilliant flash of light.

energy electron, an electron whose energy is less than nothing?

Figure 7-1 shows the energy and spin states of the Dirac electron. The upper part of the diagram contains the positive-energy states; the lower part contains the negative-energy state. A forbidden zone of width $2mc^2$—energies for which the Dirac equation has no solutions—separates the positive-energy (PE) from the negative-energy (NE) states.

A PE electron is in its lowest energy state when it is not moving; at that point a PE electron has only a rest-mass energy equal to mc^2. The faster a PE electron moves, the more energy it acquires over and above its rest mass. As a PE electron's momentum increases it moves up the energy ladder.

When a NE electron is not moving, it resides in its highest energy state with rest mass equal to $-mc^2$. Presumably, a particle with negative mass would fall up in a gravity field, but no such particle has ever been observed. As an NE electron moves faster, it actually loses energy and moves down on this ladder diagram. When you add energy to an NE electron, it slows down, a feature the NE electron shares with the legendary tachyon. Unlike the tachyon, whose energy possesses a lower bound, the NE electron can lose energy without limit, continually increasing its velocity (but never exceeding the velocity of light) as it approaches an extreme state of motion where it travels at the speed of light with an infinite negative energy.

One easy solution to the negative-energy puzzle might be to simply reject these negative solutions on the grounds that particles with such properties have never been observed. However this option is open only if we do not add electromagnetic interactions to the Dirac equation. When this interaction is put in, positive and negative solutions mix in an inextricable manner and the Dirac equation leads to a preposterous prediction: that the world will end in less than 10 billionths of a second!

This result is certainly absurd. Astronomical evidence suggests that the world is at least 10 billion years old, 25 orders of magnitude larger than Dirac's prediction. How can this theory be saved? The reason that the world explodes in an instant can be traced to those puzzling negative-energy states. Because of the electromagnetic interaction, a positive-energy electron can change into a negative-energy electron by emitting a photon of light that carries away the energy deficit between the two states. Dirac's theory predicts the time for such a transition to be about 10 billionths of a second. This is the time it would take for all the world's electrons to fall into negative (antigravity) states with the emission of an immense flash of light.

To account for the blatant existence of the world without giving up the negative-energy states, Dirac assumed that these states are already filled with electrons. These densely packed electrons—collectively called the "Dirac Sea"—are not directly observable but form the background against which everything else is observed. In Dirac's scheme the vacuum is empty of positive-energy electrons but contains a completely filled energy ocean of negative electrons. Dirac's scheme works to stabilize the electrons because an electron happens to be a type of quantum particle called a "fermion."

All quantum particles belong to one of two families depending on their intrinsic spins. Particles with integer spin (0, 1, 2, 3, . . .) are called bosons (after Indian physicist Satyendra Bose); particles with half-integer spin (1/2, 3/2, 5/2, . . .) are called fermions (after Italian physicist Enrico Fermi). The most important difference between these two types of particles is that fermions obey the Pauli exclusion principle—only one fermion can occupy a given quantum state. On the other hand, any number of bosons can occupy the same state. In a certain sense, fermions are reclusive particles—they like to be by themselves—and bosons are gregarious—they don't mind living right on top of one another.

According to the Dirac equation, electrons possess a spin of 1/2: they qualify as fermions and should obey the exclusion principle. Therefore if the negative-energy states are already filled, as Dirac assumed, then the positive-energy electrons, even in the presence of interactions mixing the two kinds of states, cannot turn into negative-energy electrons because these negative-energy states are completely occupied.

By adding energy to a negative-energy electron, however, one can change it into a positive-energy electron, lifting it out of the Dirac Sea. The minimum amount of energy that must be supplied to lift an electron out of the

Sea of Dirac is $2mc^2$—the magnitude of the energy gap that separates the two kinds of electrons. This energy can be supplied, for instance, by a highly energetic photon, carrying at least a million times more energy than a photon of visible light. It takes extraordinarily high energies to pop an electron out of the Dirac Sea.

When this energetic photon promotes an NE electron to PE status, it leaves behind an empty state in the Dirac Sea. This so-called "hole"—a sort of "bubble" in the Dirac Sea—will behave like a particle too, with properties generally opposite to the NE electron, because it represents an absence of such an electron. The NE electron had negative electric charge and negative mass (antigravity). The hole therefore must act as if it had positive charge and positive mass (ordinary gravity).

The filled Dirac Sea prevents the world from collapsing and in the process predicts the existence of a new particle—represented by a "bubble" in the sea—which can be created "out of nothing" by a sufficiently energetic photon. Because its properties are generally the opposite of electron attributes, this bubble particle is called the antielectron, or "positron." Respect for Dirac's theory grew immensely in 1932 when Carl Anderson at CalTech discovered the positron among some photographs of cosmic-ray-induced events. Today physicists have generalized Dirac's result and predict that all elementary particles come in pairs—the existence of a certain particle implies that a corresponding antiparticle also exists.

Despite the Dirac scheme's close agreement with experiment, physicists have always been uneasy about the fact that to explain the motion of just one visible electron, Dirac had to assume the existence of an infinite number of invisible electrons filling up the Dirac Sea. In the early forties, eccentric Swiss aristocrat Baron Ernst Carl Stückelberg and Princeton whiz-kid Richard Feynman independently discovered a way to solve the negative-energy puzzle without invoking an omnipresent background of antigravity electrons.

*　　　*　　　*

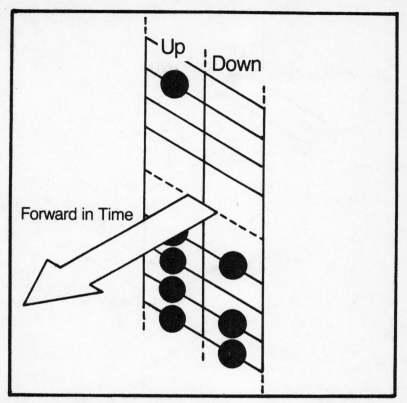

Figure 7-2. The Dirac Sea scheme for making electrons stable is illustrated. Normally all negative-energy states are filled up with electrons. The Pauli exclusion principle (only one electron allowed per state) then acts to prevent transitions from positive to negative states. Vacancies (bubbles) in the Dirac Sea look like particles with opposite properties from the absent electron. The empty hole state illustrated here acts like a positively charged, spin-up electron, called the positron or antielectron. In Dirac's scheme all electrons, no matter what their energy, travel forward in time.

The Stückelberg-Feynman (S-F) scheme exploits the ambiguity of wave equations discussed in Chapter 5—that such time-symmetric equations allow solutions that travel backward in time as well as solutions that travel forward in time. We saw there that the customary way of dealing with the backward-in-time solutions is to ignore them on the

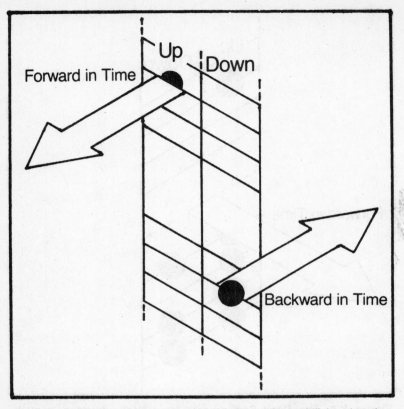

Figure 7-3. The Stückelberg-Feynman scheme for stabilizing the electron. Positive-energy states move forward in time; negative-energy states move backward in time. Positive states cannot transit into negative states because no negative states in nature are moving in the positive time direction. We humans, who are compelled for unknown psychological reasons to see time flowing in only one direction, experience a backward-in-time electron as a positive-energy electron with reversed attributes. In our world the negative-energy, time-reversed state depicted here looks like a positively charged spin-up positron, behaving exactly like the bubble state in Figure 7-2.

grounds that nature has apparently chosen not to employ these time-reversed options. Like most other equations in physics, the Dirac equation is time-symmetric and has two solutions. Stückelberg and Feynman proposed that the world is such that both these solutions exist in nature in particular combinations: All positive-energy solutions run forward in time, while the troublesome negative-energy solutions run backward in time. Solutions of the other kind—positive-energy/backward in time, negative-energy/forward in time are forbidden—are options nature has chosen not to use.

This S-F scheme works to prevent the world from collapsing by forbidding positive-energy electrons from turning into negative-energy electrons. Such an event is impossible because it would involve a PE electron traveling forward in time turning into an NE electron traveling forward in time, but the S-F scheme does not contain any NE electrons traveling forward in time—the only kind of NE electron allowed is one that travels backward in time. To humans whose experience always runs forward in time, an NE electron going backward in time looks like an electron with all its signs reversed—positive mass, positive charge, positive energy, spin pointing in the opposite direction—in short all the properties observed for the positron. A possible history for a PE electron is to change into a NE electron going backward in time. Viewed from our forward-time perspective, this process looks like an electron and a positron both traveling forward in time which meet and vanish in a flash of energy—a matter-antimatter annihilation event.

In Dirac's scheme (before he added his filled sea), PE electrons were unstable in vacuum because nothing prevented them from turning into NE electrons. In the S-F scheme, PE electrons are stable in a vacuum but unstable against positron collision, but since positrons are relatively rare in our part of the universe, electrons will last virtually forever here.

Feynman has shown that the S-F scheme is mathematically equivalent to the filled Dirac Sea, both schemes give the same answers to all calculations. If both schemes are experimentally equivalent how can we ever hope to know if positrons are really electrons traveling "backward in time"?

In the laboratory, a positron reveals no trace of any exotic time-reversed behavior. It acts precisely like an electron that just happens to have positive charge. Despite the efficacy of the S-F formalism—which is used today for calculating the behavior of all particles, not just electrons—no one has ever figured out any way of using positrons to send signals into the past. Also it is only a historical accident that the Dirac equation was written for electrons. It could just as well have been written for positrons. In that case, the S-F scheme would have positrons running forward in time while electrons run backward. There is, of course, no experimental difference between these two schemes. The present convention is to call the electron (negatively charged particle) the particle that travels forward in time, and to call the positron (positively charged particle) the antiparticle that travels backward in time, but the opposite convention would also work. Thus not only can we not demonstrate directly that the positron is really an electron traveling backward in time, we cannot even be sure that it is not the electron that is the one traveling against the timeflow direction that humans consider ordinary.

There is some evidence, however, that the S-F scheme is a more realistic way of looking at the electron than the Dirac Sea scheme despite the fact that both methods predict the same outcomes for experiments. The Dirac scheme will only work for fermions. For bosons, which do not obey the Pauli exclusion principle, filling up the NE states with invisible particles will not prevent the PE particles from tumbling down the energy ladder into those NE states. Bosons (such as the spin-zero pion and the spin-1 "weakons") obey relativistic equations that like Dirac's equation possess both positive- and negative-energy solutions. Both bosons and antibosons are observed in the laboratory. For bosons the S-F scheme is the only method available to insure particle stability, suggesting that this scheme's backward-in-time way of viewing antiparticles is closer to the truth of the matter than Dirac's vision of an omnipresent ocean of antigravity states.

If the S-F scheme contains a clue concerning how to build time machines, nobody has yet deciphered it. In all experiments, both particles and antiparticles look perfectly ordinary—like positive-energy particles traveling forward in time. No hint ever emerges in the real world that one class of particles—we are not sure which one—might be really moving in the opposite direction to our psychological time sense.

Neutral Kaons

A convenient way to describe certain general features of the world is in terms of "symmetry." To say that a symmetry (of type X, for instance) exists means that the observation of property A necessarily implies the existence of property B, its X-symmetric partner. For instance, in the last section, we learned that the existence of an elementary particle in nature automatically entails the existence of a corresponding antiparticle with the same mass but opposite charge—a situation physicists call "C symmetry."

Another conceivable symmetry (call it G symmetry) might be that the existence of a particle of mass m always implies the existence of an antigravity partner of mass $-m$. We appear to live in a universe where C is a good symmetry (except, as we shall see, in certain rare reactions) but G is not a symmetry of this world at all. As far as we can tell, the Universe contains no antigravity particles of any kind. One of the big open questions in physics today is whether the Universe is electric-magnetic symmetric, in the sense that the observed existence of electric charges implies the existence of magnetic charges—so-called "magnetic monopoles."

Several decades of searching for a monopole have turned up nothing. However two recent experiments, one at Stanford in 1981 and one at Imperial College, London in 1986, have apparently yielded one monopole each. If these rare events continue to accumulate, we will soon be able to say that nature possesses a pleasant symmetry between her electric and magnetic interactions.

The existence of a particular symmetry implies—all other things being equal—that the number of particles should be the same as the number of their symmetry partners. If particle/symmetry partner equality is not observed some special mechanism must be invoked to explain this inequity—why is one particle produced in preference to the other? For instance, experiments show that the Universe is made mainly of particles rather than antiparticles. The stars, the galaxies, and the interstellar medium are almost wholly constituted of normal matter not antimatter. The rarity of antimatter poses one of the major problems of modern cosmology: If the laws of physics are indeed *C*-symmetric, where did all the antimatter go? If the laws of physics are found to be electric-magnetic symmetric by the discovery of more monopoles, a similar mystery emerges. The scarcity of magnetic charges compared to the ubiquity of electric charges raises the question: Where are the rest of the monopoles?

Any Big Bang model of the creation of the Universe must account for the preferential creation of normal matter and electric charge rather than antimatter and magnetic charge. Soviet physicist Andrei Sakharov has developed a set of minimum theoretical conditions that must have been present in the primordial fireball in order that it produce the present asymmetric population of elementary particles. All current models of the Big Bang incorporate some form of the Sakharov conditions.

A symmetry that might prove of interest to time travelers is the so-called *T*-symmetry (or time-reversal symmetry), which asserts that for any interaction between elementary particles, the time-reversed reaction—a movie of the original reaction run backward—is also possible. Moreover, the probability (or "reaction strength") of the two reactions is precisely the same. For example, in a *T*-symmetric world, if particles *A* + *B* combine to make particles *C* + *D*, then particles *C* + *D* will combine with exactly the same probability to produce particles *A* + *B*.

Symmetries like *T* require not only the existence of particle-partner pairs but also particle-partner reactions. For example, *C* symmetry requires not only that matter and

antimatter exist but also that any reaction that matter particles undergo implies the existence of a corresponding antimatter reaction of precisely equal strength—a reaction in
which all matter particles are replaced by their anti-matter
counterparts. Thus the existence of the reaction $A + B$ ʟ C
$+ D$ implies the existence of the reaction $\bar{A} + \bar{B}$ ʟ $\bar{C} + \bar{D}$
(where the overbars indicate antiparticles) if C symmetry is
valid.

Interest in symmetry principles skyrocketed in 1957 when
Chinese-American physicists Tsung Dao Lee and Chen Ning
Yang proposed, and experiments subsequently verified, that
the parity symmetry was violated in weak interactions. Parity
(P symmetry), sometimes called "mirror-reflection symmetry," requires that for any particle/reaction in nature, the
mirror image of that particle/reaction must also exist. Since
a mirror changes a right hand into a left hand, the observed
violation of parity symmetry means that certain "left-handed"
particles and reactions exist but not the corresponding right-
handed particles and reactions. But what exactly is a "left-
handed particle"?

All elementary particles possess a property called "spin,"
which we might visualize as a certain axis around which the
particle executes a rotary motion. Spin, along with momentum, energy, and position, is a quantum attribute. As with
all such attributes, the visualization of spin as an actual
rotation should be taken with a grain of salt. For some
purposes such a picture is adequate but it cannot be pushed
too far.

For most particles the spin axis can point in any direction, but for a massless particle—such as the photon or
neutrino—the spin axis is constrained to lie along the direction of the particle's linear motion. Look along the particle's path as it goes away from you. If the particle is
massless you will also be looking along the spin axis. If the
particle has mass, consider only that component of spin
that does point along its trajectory. Now if you see the
retreating particle turning clockwise, the particle is "right-

handed"; if it rotates counterclockwise as it goes away from you, it's a "left-handed" particle.

The neutron is an unstable particle that decays (with a half-life of about 15 minutes) to a proton, an electron, and an antineutrino:

$$n \rightarrow p + e + \bar{\nu}$$

This reaction violates P symmetry because the antineutrinos from this reaction are always measured to be right-handed. In fact, every reaction that makes antineutrinos always makes right-handed ones, never any left-handed ones. As a result, the Universe, as far as we can tell, contains not a single left-handed antineutrino, a gross violation of P symmetry. Neutrinos are only produced in the so-called "weak" interactions. Nature's other three basic interactions—strong, electromagnetic, and gravitational—seem to be perfectly mirror-symmetric.

Shortly after the discovery of parity violation, physicists learned that the weak interaction also violates C symmetry. For instance, the C-symmetric partner to the neutron decay:

$$n \rightarrow p + e + \bar{\nu} \qquad \text{(right-handed } \bar{\nu}\text{)}$$

should be

$$\bar{n} \rightarrow \bar{p} + \bar{e} + \nu \qquad \text{(right-handed } \nu\text{)}$$

That is, an antineutron should decay—with the same 15-minute half-life as the neutron—into an antiproton, an anti-electron, and a right-handed normal-matter neutrino. This C-symmetric reaction is never observed. The antineutron does indeed decay—with the correct half-life—but the neutrino is left-handed rather than right. In fact, all the neutrinos in the world appear to be left-handed; there is no such thing as a right-handed neutrino.

To restore a sense of symmetry to the weak interactions, Soviet physicist Lev Landau and others proposed that these interactions obeyed a symmetry they called CP. Under CP symmetry, changing all particles in the reaction to antiparticles and also reflecting the reaction in a mirror yields another reaction that will be observed in nature. In this proposal weak interactions are not symmetric under C or P separately but are symmetric under their combination. Following this new rule, the CP partner to neutron decay is just:

$$\bar{n} \rightarrow \bar{p} + \bar{e} + \nu \qquad \text{(left-handed } \nu\text{)}$$

This is precisely the kind of antineutron decay that is observed. Other weak interactions in which the spins of the particles could be measured were also found to conform to this combined C and P symmetry. For a while it looked as if all interactions except the weak ones were symmetric under C and P alone; the weak interactions were symmetric under C and P together.

A powerful argument in favor of CP symmetry is the CPT theorem of Wolfgang Pauli and Fritz Villars. In 1949, these physicists showed that relativistic quantum field theory, the general framework within which all particle physics is interpreted today, is necessarily symmetric with respect to the combined operations of C, P, and T. If a reaction violates CP symmetry, it must also violate T to the same extent in order that CPT remain a good symmetry. A reaction that violates T symmetry has a different strength depending on whether the reaction is going forward or backward in time. In a universe that violates T symmetry, it is possible to distinguish, on the level of elementary particles, a movie running forward from one running backward.

In the macroscopic world, it is easy to tell if a movie is running backward, because, for instance, eggs never unscramble themselves in the forward-time world. However this time asymmetry on the macroscopic level is attributed to the statistical rule that unhindered events move from less probable to more probable configurations, plus the contingent fact that much of the everyday world—the unscrambled egg, for example—happens to be in exceedingly improbable states.

In the microscopic world, on the other hand, things are much simpler. The fundamental events from which the world is constructed are not in improbable configurations. Hence these events are expected to be perfectly time-symmetric. All tests we have been able to carry out do indeed validate the T symmetry of elementary particle reactions.

In light of the CPT theorem and good tests of CP and T symmetry, it came as somewhat of a surprise when James Cronin, Val Fitch, and their collaborators at Brookhaven showed in 1964 that the weak decay of a certain obscure particle—the neutral kaon (pronounced "kay-on")—violated CP symmetry to the extent of a few parts per thousand.

Two varieties of neutral kaon exist in nature, called "K-long (K_L)" and "K-short (K_S)". These names refer to the kaon's lifetimes, which differ by a factor of 500. If these two kaons possessed identical decay channels, their lifetimes would be the same. Their lifetimes are different because of CP symmetry which permits the K_S to decay into two pions but forbids the K_L to decay in this way:

$$K_S \rightarrow \pi_0 + \pi \quad \text{(two neutral pions)}$$
$$K_S \rightarrow \pi^+ + \pi^- \quad \text{(two charged pions)}$$

The K_L is permitted by CP symmetry to decay into three particles, a pion, a lepton (electron or muon), and a neutrino (or antineutrino):

$$K_L \rightarrow \pi^- + l^+ + \nu \quad \text{(positive lepton)}$$
$$K_L \rightarrow \pi^+ + l^+ + \bar{\nu} \quad \text{(negative lepton)}$$

CP symmetry permits this three-body decay and also requires that the rate of production of positive leptons be exactly equal to the rate of production of negative leptons.

If CP symmetry is valid, it constrains the decay of the K_L particle in two ways: (1) K_L must never decay into two pions and (2) the allowed three-body decay of K_L must produce equal numbers of positive and negative leptons.

The results of the Cronin-Fitch experiment violate both these expectations: (1) K_L's are found to decay a small percentage of the time into two pions (two events out of every thousand) and (2) in three-body decays, positive leptons predominate by about three events per thousand (0.3 percent effect). Hence *CP* symmetry demonstrably fails to describe neutral kaon decay by a tiny but undeniable amount. Only neutral kaons show this effect. All other reactions within the limits of measurement accuracy display perfect *CP* symmetry.

The *CPT* theorem requires that any reaction violating CP symmetry must also violate *T* symmetry to the same extent, in order that an overall *CPT* symmetry be maintained. Therefore the production of neutral kaons from pions should proceed at a slightly different rate from the reverse reaction —the decay of kaons into pions. So far this conclusion remains only theoretical because direct experiments to test the *T* symmetry of kaon production are extremely difficult to carry out. Tests of *T* symmetry in other weak interaction processes that are more experimentally accessible (such as neutron decay) have not revealed any violations. These measurements, however, have not yet reached the 0.1 percent level of accuracy at which *T* violation is expected to manifest. At present, violation of *T* symmetry has not been directly demonstrated but is inferred to occur in the kaon reaction because *CP* is violated and the *CPT* theorem is believed to be valid.

The origin of CP violation (and its accompanying *T* violation) in nature is completely mysterious. It cannot be understood within the conventional four-force framework of physics which has prompted some physicists to postulate the existence of a fifth ("superweak") force whose sole function is to account for *CP*-symmetry-breaking in neutral kaon reactions. To account for the good *CP* symmetry observed everywhere else, the strength of this new force is tailored so that its effects in other reactions are unobservably small.

Minuscule and mysterious as it is, this tiny *CP* symmetry violation may have played a big part in determining the structure of the early Universe. In the orgy of particle creation following the Big Bang, particles and antiparticles would be created with equal probability unless some sort of built-in asymmetry biased the Universe in favor of matter rather than antimatter. The *CP* violation in neutral kaon decay is just such a bias and indeed *CP* violation is one of the Sakharov conditions for producing a matter-dominated Universe out of the Big Bang. Without these constraints, stars and galaxies (all made of matter) would not have formed, only a matter-antimatter soup quickly evaporating in a burst of radiant energy. Such a universe would consist mainly of radiation, with a few lucky particles and antiparticles floating far apart from one another. Thus our very existence as solid bodies may in some sense be dependent on the tiny wisp of *CP* asymmetry revealed in the Cronin-Fitch experiment.

While the Universe's tiny *T*-symmetry does not seem to offer a mechanism for time travel, it does allow us to distinguish an absolute temporal direction. Certain elementary particle reactions proceed at a different rate depending on the direction of time's arrow. Thus in the far future, the K_L decay might serve as a convenient signpost for time travelers.

After exploring the multiple universe in your model *CPT* tempocruiser, you return to your garage. At least you think it's your garage. Everything looks all right outside. The world is made of ordinary matter, rather than antimatter. Neutrinos are all left-handed, not right-handed. But before you call it home and decide to disembark, ring up the physics library for data on K_L decay. Look up the dominant three-body lepton and ask: What's your sign? If positive leptons predominate slightly in these decays, you are in a universe very much like the one you left. You might as well call it home. However if the dominant leptons are negative, get out! You have landed in a time-reversed universe where time flows in a direction opposite to our own.

CHAPTER 8

The Quantum Connection:
A Superluminal Shortcut
In Configuration Space

"It can't be wrong. The mathematics of it is ironclad. Every particle in my body and yours, my dear, is communicating faster than the speed of light with every particle of everybody else who was at Mary Wildeblood's party tonight . . . faster than the speed of light, mind."

"Cripes. That's spooky when you say it that way."

"Spooky indeed. It is identical with what we anthropologists call the Law of Contagion, which is the um savage superstition ha-ha that if you can get hold of the hair or fingernails of somebody you can control him at a distance."

—Robert Anton Wilson, *Schrödinger's Cat*, 1979

No branch of science is more suggestive of possibilities for magically transcending the limits of space-time than quantum theory. In the strange world of the quantum, a particle can vanish without a trace (quantum annihilation), or come into existence out of nowhere (quantum creation), move from location *A* to location *B* without being in between (quantum tunneling), or instantly flip from one state of being to another (quantum jumping). So far no one in science fiction or science fact has attempted to use particle annihilation and creation, quantum tunneling or quantum jumping for faster-than-light communication. Most recent proposals to use quantum mechanics for superluminal communication focus on the so-called "quantum connection"—the notion that once two quantum systems have briefly interacted, they remain in some sense forever connected by an instantaneous link—a link whose effects are undiminished by interposed shielding or distance. Physicists' belief in the reality of the quantum connection has been strengthened by the advent of Bell's theorem, which proves that any model of how events happen in the world that leaves out this superluminal link will fail to explain certain simple quantum facts.

The quantum connection was discovered in 1935 by Austrian physicist Erwin Schrödinger who considered this residual instant link between distant particles the most peculiar feature of quantum theory. Schrödinger showed that when two quantum particles, A and B, interact briefly by conventional means such as electric or magnetic forces, then move far apart beyond the range of the interaction, the mathematical probability waves that represent the particles A and B do not separate cleanly. Instead these waves remain "phase entangled," stuck together in such a way that when the wave that represents particle A is changed, a corresponding change occurs instantly in the wave that represents particle B.

The fact that the actual systems can separate while the quantum waves that represent the systems do not separate is possible because quantum waves do not dwell in ordinary three-dimensional space, but in a larger multidimensional space called "configuration space" which possesses three dimensions for each particle. The quantum wave for a two-particle system, for instance, moves about in a six-dimensional space.

This superluminal quantum connection is not mediated by a force field like all other physical interactions but seems to work more like voodoo: An action committed on particle A instantly affects particle B because during their brief conventional interaction, particle B left a part of itself with particle A, a part with which it is still in contact. Among the qualities that characterize waves such as frequency and amplitude is a quality called "phase." Like the phases of the moon, a wave's phase tells in what part of its cycle the wave currently resides. The "parts" that A and B leave in one another to establish the quantum connection are pieces of one another's phases. Because nothing actually stretches between A and B to mediate their linkage, this connection cannot be shielded by interposed matter and its strength does not diminish with distance. The quantum connection is, in short, unmediated, unmitigated, and immediate.

Schrödinger and other physicists were careful to point out that despite the presence of this superluminal phase entanglement in quantum theory, no superluminally connected quantum facts had ever been observed. In 1978 Berkeley physicist Philippe Eberhard showed (Eberhard's proof) that although superluminal effects do indeed occur at certain stages of the quantum calculations, these effects always cancel themselves out before the final results appear, so that all quantum predictions must, in fact, obey the Einstein limit. For good experimental and theoretical reasons the quantum connection was regarded for more than 30 years as a mere theoretical artifact with no real practical consequences. Most physicists were convinced that one could no more use the quantum connection for FTL signaling, than one could use the International Date Line for time travel. To those who had completely dismissed the notion of real FTL connections in the quantum world, the theorem of John Bell, which demonstrates the necessity of FTL connections, came as a great shock.

Bell's Theorem

Bell's theorem shows that the quantum connection is not a mere theoretical artifact but corresponds to a real superluminal link that actually exists between any two phase-entangled systems. Bell proved his theorem in a negative sort of way, by assuming that such real links did not exist, and showing that the experimental consequences of such a "locality assumption" are contrary to what is observed in the laboratory. The experiments are subtle, they do not show directly the presence of superluminal connections, but demonstrate instead the futility of trying to explain the experiments by subluminal connections alone. Bell's proof, published without fanfare in an obscure and short-lived journal called *Physics*, shows that any conceptual model of how individual quantum events are produced that does not incorporate FTL connections will fail to explain the results of certain simple physics experiments. These experiments, carried out by John Clauser at the University of California at Berkeley and Alain Aspect at the University of Paris, are indirect but indisputable indications of the presence of a real superluminal connection in nature. Bell's theorem says not merely that superluminal connections are possible, but that they are necessary to make our kind of Universe work.

But if Eberhard proved that all quantum facts must be slower-than-light, how was it possible for Bell to show that the same quantum facts must be connected faster-than-light? Bell and Eberhard can both be right because they are each talking about a different aspect of a quantum measurement. All quantum measurements when scrutinized at their finest level of resolution consists of tiny particlelike events called "quanta," or "quantum jumps"—flashes of light on a phosphor screen, for instance; or a bubble, spark, or click in a particle detector; the blackening of a silver grain in a photographic emulsion; or the sudden excitation of a light-sensitive molecule in your eye. The world when looked at closely appears to be made of little dots, much like color photos in a magazine. The first law of quantum theory is that these quantum jumps occur completely at random—no theory, quantum or otherwise, can predict where or when the next light-induced flash will occur on your retina.

The job of quantum theory is not to predict when and where the next quantum jump will occur—no theory can do that because these jumps are truly random—but to describe patterns of quantum jumps. When a large number of quanta have been recorded, they form a regular pattern, the ultimate quantal pattern being the world we see around us. Quantum theory has been enormously successful at this task of pattern prediction. In its fifty years of existence, physicists have applied quantum theory to a vast range of physical phenomena from predicting the confinement of quarks in a nucleon to describing the types of particles created in the Big Bang and this theory has never made a false prediction.

A quantum jump may be compared to a single roll of a pair of dice. The outcome of a single throw is unpredictable, this is what makes dice a game of chance. Although each dice roll has a random outcome, after a large number of rolls a pattern emerges, a pattern that reflects the dice odds on which the payoff structure of the game of craps is based. Likewise, quantum theory is unable to explain single quantum events but only predicts the "dice odds" that govern the fundamental processes that run the world, an odds structure that is revealed only when large numbers of quanta come into play.

Eberhard's proof applies to the quantum patterns—the "dice odds" of the Universe. Eberhard's proof guarantees that large-scale quantum patterns will never be observed to be connected faster than light. Bell's theorem, on the other hand, applies to the individual quantum events themselves, and proves that these little quantum jumps must be connected faster than light. An uneasy truce exists among Bell theorem, Eberhard's proof, and the COP rule of special relativity. Although they make opposite claims, Bell's theorem can coexist with Eberhard's proof because they each refer to different aspects of a quantum measurement. Bell's theorem, which necessitates a real superluminal connection in nature, can also coexist with the COP rule forbidding all superluminal connections that can be used for signaling, because these Bell-mandated FTL jumps occur in an utterly random manner. Since a signal always involves the transmission of some intelligible pattern, and since random events are by their very nature patternless, random events are totally useless as a message medium.

A quantum measurement might be compared to a television picture tube whose red-, green-, and blue-colored dots flash on and off at random rather than in the regular raster-scan mode of a conventional set. The randomly flashing dots represent quantum jumps and the picture represents the quantum pattern. The experimenter's role is to choose what measurement he will perform by turning the channel selector. The role of quantum theory is to predict what picture will appear on the screen for each position of the dial.

To simulate the quantum connection, we add another television. Imagine two television sets, a blue Sony and a green RCA. These sets bump into one another as they are placed on the shelf in the store. As their enclosures touch they become quantum-correlated, the atoms that make up the blue set become phase entangled with the atoms of the green. We imagine that this brief interaction is strong enough to connect the insides of each television as well as their cabinets.

The televisions are sold; Max buys the blue one and takes it to Miami, and Maxine takes the green one to Minneapolis. The televisions cease to interact by conventional means. When Max and Maxine are both watching channel 1, they do not notice that thanks to the quantum connection the sequence of colored dots that make up their pictures is exactly the same. If Max would pay attention to the dots rather than the picture, he might see a sequence:

$$R\ G\ R\ R\ B\ G\ B\ G \dots$$

Because of the quantum connection Maxine would see an identical sequence of dots that appear on her screen at exactly the same time as Max's dots, that is without the usual time delay associated with the distance between Miami and Minneapolis. Although these sequences are quantum connected, they are also random, so no message is being carried by their FTL correlation.

Now Max switches channels—analogous to a physicist's choice to make another kind of quantum measurement on his system—and a different picture begins to form, dot by dot, on Max's screen. Maxine's picture, of course, does not change when Max changes channels, in agreement with Eberhard's proof that quantum patterns are not linked by the quantum connection. However, Bell's theorem demands that not only do Max's random dot sequences alter when Max changes channels but Maxine's dot sequence alters as well. Furthermore, her sequence responds instantly, without delay, to Max's distant channel change.

When Max changes channels, his television picture as well as the "random" dance of his dots change their behavior. Because of the assumed quantum connection linking the two television sets, Max's distant channel change alters the "random" dance of Maxine's dots—they flash on and off in a different sequence than if Max had continued to watch the same channel—but Maxine's television picture does not change. If we could detect this Max-induced change in Maxine's dot dance, we could signal faster-than-light using the quantum connection.

But how could Maxine's purported sequence change ever be detected? One random sequence looks pretty much like any other random sequence. It is this indistinguishability of random sequences that prevents the use of the quantum connection as a superluminal communication channel. Superluminal messages can be sent—Maxine's sequences undoubtedly change, we are assured by Bell's theorem, when Max flicks his channel selector—but because such messages cannot be decoded, Maxine can never become aware of this instant change. The utter inscrutability of random sequences has been aptly described in a phrase from a crystallography textbook: "Although there are many kinds of order, there is only one kind of randomness."

The random occurrence of quantum jumps, which seems on the face of it to block all attempts to use the quantum connection for superluminal signaling, has not prevented some people from trying to devise ingenious schemes to decode the instantaneous messages locked inside these unpredictable sequences of quantum jumps.

Most superluminal communication schemes that attempt to exploit the quantum connection are based on a particular two-body quantum system called the EPR experiment after Albert Einstein, Boris Podolsky, and Nathan Rosen, who used this experiment in 1935 to illustrate some unusual consequences of the quantum connection. The EPR experiment is one of the simplest examples of quantum-connectedness but it is complex enough to illustrate all the important features of this putative superluminal link.

The EPR experiment consists of a light source made up of excited calcium atoms, similar to the excited neon atoms in a neon sign. Unlike neon, which gives off one photon per atom, each calcium atom produces two photons (blue and green) which fly off from their source in opposite directions at the speed of light. These photons travel to distant detection sites where their polarization is analyzed with a calcite crystal and a pair of photon counters.

The EPR experiment employs two calcite crystals to measure the polarization of the blue light beam and the polarization for the green light beam. The intent of this experiment is to measure the extent to which the polarizations of the blue and green photons have become correlated by virtue of these two photons having once been together inside the same calcium atom.

In the quantum picture light possesses both a particle and a wave aspect. Light's polarization in the wave picture refers to the direction in which the wave is vibrating; in the particle picture polarization is associated with the particle's spin: The polarization of a photon, for instance, may be visualized as a little ball with an arrow sticking through it indicating the photon's polarization direction. Only photons polarized at right angles to the photon's line of flight are important here and we will limit our attention to the four polarization directions labeled H, V, D, and S.

H and V stand for horizontally and vertically polarized photons, respectively. The arrow associated with a horizontally polarized photon points in the horizontal direction at right angles to the photon's direction of travel. The vertical photon's arrow points at right angles to both the horizontal arrow and the photon's line of flight. D and S stand for diagonal and slant polarized photons, respectively. The diagonal direction lies to the right of vertical halfway between the H and V directions. The slant direction lies to the left of vertical, halfway between the *H* and *V* directions.

In the EPR experiment, when a pair of blue and green photons emerge from the calcium-vapor source, they are completely unpolarized. No definite direction can be as-

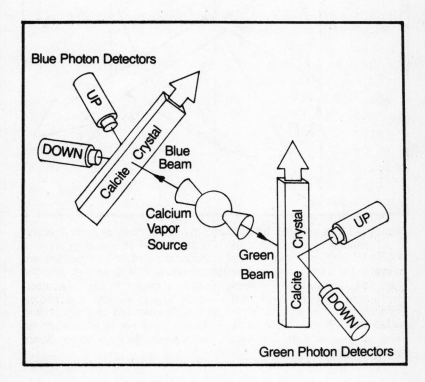

Figure 8-1. The Einstein-Podolsky-Rosen (EPR) experiment is diagramed. A calcium-vapor light source emits a quantum-correlated pair of blue and green photons that travel in opposite directions at light speed to two calcite crystals where they are deflected into up or down photon detectors depending on their polarization.

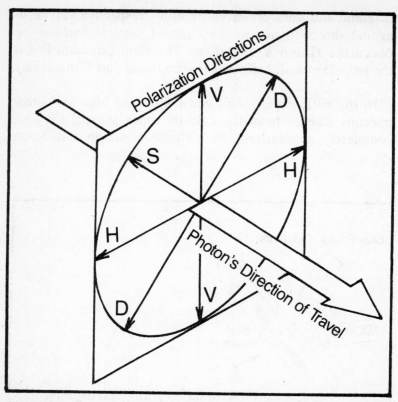

Figure 8-2. Photons can be polarized in any direction, or in no direction at all (unpolarized or *U* photons). For the EPR FTL signaling scheme described here, only two pairs of orthogonal polarization directions are relevant: the vertical (*V*) and horizontal (*H*) pair of directions, and the diagonal (*D*) and slant (*S*) directions. When a photon of any polarization encounters a calcite crystal whose optic axis is vertical, that photon must change into either a *V* photon (which goes up) or an *H* photon (which goes down). Likewise, a diagonal calcite forces photons to be either *D* polarized or *S* polarized, whereupon they go up or down, respectively.

signed to such photons, which we will call "*U* photons." After emission, a *U* photon remains in this indecisive polarization state from the time it leaves the light source until its encounter with the calcite detector system, whereupon the *U* photon acquires a definite polarization that depends on the setting of the calcite crystal.

Calcite is a transparent mineral discovered in Isaac Newton's day. This crystal is "birefringent," which means that a photon that happens to encounter a chunk of calcite can travel in one of two directions through the crystal. In the seventeenth century these two light paths were called "the ordinary ray" and "the extraordinary ray," but I call these paths "up" and "down" for short. Each photon that goes into a birefringent mineral must make a decision either to go up or down, a decision that depends both on the photon's polarization direction and on the direction of the calcite crystal's optic axis.

When the photon enters the crystal, it may have any polarization whatsoever, but when it comes out, its polarization is forced to point along one of two directions, either along the crystal's optic axis if the photon goes up, or at right angles to the crystal axis if the photon goes down. For instance, if the calcite axis is set in the vertical direction, all photons that go up become vertically polarized; those that go down are all horizontally polarized.

What path does a polarized photon take when it runs into a calcite crystal? Physicists answered this question two hundred years ago by shining light of various polarizations into calcite crystals oriented along various directions.

If the photon is polarized in the same direction as the calcite's axis, or at right angles to this axis, the photon has no choice. The same-direction photons always go up; the right-angled photons always go down. When a V photon, for instance, encounters a vertical calcite, the photon must go up; when a horizontal photon meets a vertical calcite, it must go down. Photons like these whose behavior vis à vis a calcite crystal is completely predictable, I will call "perfectly infallible photons," or PIPs.

When a photon, polarized in a direction (such as S or D) halfway between vertical and horizontal, encounters a vertical calcite, it has a 50/50 chance of going up or down. This decision is quantum random, meaning that nothing in the laws of physics can predict in what direction that photon will go. Polarized photons that find themselves confronted with two equally likely outcomes, I call "perfectly optional photons," or POPs. In all situations discussed here, a polarized photon encountering a calcite crystal will be either a PIP (determinate) or a POP (random) kind of photon.

The behavior of an unpolarized, U photon, when it encounters a calcite crystal is discussed later in the context of a particular proposal to use the EPR experiment as the crucial component of a time machine.

How to Build a Time Machine

Two kinds of time machines are conceivable: one that sends people back in time and another that sends only information back into the past. If you really need to personally witness the building of the pyramids or the dedication of the Taj Mahal, then you are advised to wait for the development of spacewarp technology, one of the few FTL loopholes with cargo-carrying capacity. The quantum connection, in common with most other time-travel schemes discussed in this book, would be suitable, if it could be made to work, only for the transmission of information, not for the transfer of the physical bodies back into the past.

The ability to send information into the past is no small achievement, since such an ability would allow you to establish an effective presence in the past, if not an actual presence. To be effectively present anywhere (or anywhen), it is sufficient to be able to sense the situation and to carry out intentional acts. We already know how to sense the past via the examination of records. All that we lack is a way to carry out acts decided now in a time that is previous to the present. The ability to send information backward in time could achieve this goal because this information could be used by a colleague, or by some robotic mechanism acting by proxy in the past, to fulfill our present wishes. To be effectively present in the past, it is necessary only to have a good set of records, and to possess the ability to signal backward in time.

Special relativity shows us, as outlined in Chapter 3, that

to signal backward in time, it is only necessary to be able to signal faster than light in two appropriately moving reference frames. Now I will show how the EPR experiment can be used to signal faster than light if a certain kind of polarization measurement can be carried out.

To operate the EPR experiment as an FTL signaling system, the calcium-vapor light source is positioned at a location equidistant from sender and receiver. This source acts as a kind of photon lighthouse, shining a blue beam in one direction to the sender, and a green beam to the receiver in the opposite direction. The sender encodes his intended message as a sequence of binary bits (ones and zeros) which he inserts into the EPR apparatus as a corresponding sequence of blue calcite moves. To send a "one," he sets the blue calcite vertically; to send a "zero," he sets his calcite diagonally. The receiver does not change her green calcite's setting, keeping it oriented vertically at all times.

To demonstrate the faster-than-light potential of the EPR experiment in the most dramatic manner, Maxine moves her green (receiver) calcite out to the vicinity of Mizar, a bright star in the Big Dipper, almost 100 light-years (500 trillion miles) from downtown Miami where Max has located his blue (sender) calcite crystal. The calcium-vapor light source is repositioned halfway between the two twins, and emits pairs of unpolarized (U type) photons—blue photons to Max in Miami, green photons to Maxine on Mizar—which reach their destinations at approximately the same time.

The action of U photons on a calcite crystal is easy to describe: A U photon has a 50/50 chance of going up or down no matter how the calcite is oriented. Furthermore, if the U photon goes up, it changes from its unpolarized U state into a completely polarized photon whose polarization direction is the same as the direction of the calcite axis. If the U photon goes down, on the other hand, it becomes polarized in a direction at a right angles to the calcite's axis.

For instance, if Max decides to send a "one" and sets his blue calcite vertically, all U photons that strike his calcite and go up become V photons; all U photons that happen to go down turn into H photons.

What makes the EPR experiment potentially useful for FTL signaling is what happens to Maxine's green photon when Max's corresponding blue photon turns from a U to a V photon inside Max's blue calcite crystal. Because the photons were once together, the blue and green photons are linked by a quantum connection. In the case of the calcium-vapor source this quantum connection is as strong as can be, photon pairs emitted by calcium vapor are 100% correlated. As a consequence of this strong quantum connection, when the blue photon changes from a U to a V, its green partner 100 light-years away changes instantly from a U photon to a V photon. This instantaneous response of a green photon to a distant blue photon's change provides the mechanism for sending superluminal messages via the quantum connection.

Since Maxine on Mizar has set her green calcite vertically, her changed photon (U into V) will go up with 100 percent reliability. Max's act in Miami—setting his blue calcite in the vertical direction—and his subsequent detection of an up photon, has resulted in a change in Maxine's green photon, from a U photon with a 50/50 chance of going up or down to a V photon with a 100 percent chance of going up. In other words, Max's action has changed Maxine's former POP (random) photon into a PIP (determinate) photon.

If Max's blue photon goes down instead of up, it changes from a U photon to an H photon. At that instant, Maxine's distant U photon also changes from a U photon into an H photon. Since Maxine's calcite is set vertically, this H photon will go down with 100 percent inevitability. Once again Max's distant action has instantly changed Maxine's former POP photon into a PIP photon.

When Max sets his calcite vertically he has no control over whether the blue U photon goes up or down, whether his U photon becomes V- or H-polarized. However no

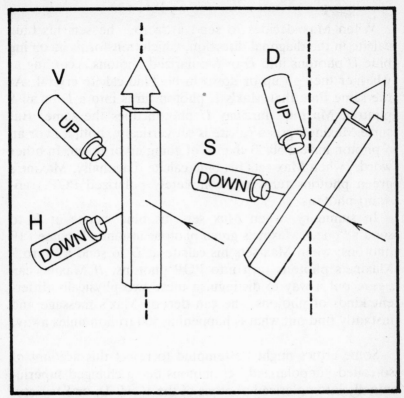

Figure 8-3. Diagram shows how to send a superluminal signal. When Max sets his blue calcite vertically, the formerly unpolarized blue photons that strike his calcite change into either *H* or *V* photons. The quantum connection instantly compels Maxine's distant unpolarized green photons to assume the same polarization. On the other hand, when Max sets his calcite diagonally, his photons change into either *D* or *S*; the distant green photons instantly assume the same polarization. Max's message—instantly transmitted to Maxine via the quantum connection— consists of a series of photons whose polarizations alternate in an intentional way between the *H-V* and the *D-S* directions. Maxine can decode Max's FTL message if she can determine, in a single measurement, the altered polarization of her green photons.

matter which kind of polarized photon Max's blue *U* becomes, Maxine's green photon instantly assumes the same polarization, changing, in either case, from a POP photon to a PIP photon with reference to her vertical green calcite.

* * *

When Max decides to send a "zero," he sets his blue calcite in the diagonal direction, which transforms incoming blue U photons into D or S polarized photons according to whether they go up or down in his blue calcite crystal. At the same time that Max's U photon turns into a D or an S photon, Maxine's faraway U photon does the same. But since Maxine's green calcite is set vertically, both a D or an S photon have a 50/50 chance of going up or down. In other words, when Max sets his blue calcite diagonally, Maxine's green photons turn into completely polarized POP (random) photons.

In summary, when Max sets his blue calcite at V to send a "one," Maxine's green photons instantly become PIP photons; when Max sets his calcite at D to send a "zero," Maxine's photons turn into POP photons. *If* Maxine can figure out a way to distinguish these two physically different kinds of photons, she can decode Max's message and instantly find out what is happening 500 trillion miles away.

Some critics might be tempted to reject this account of so-called "unpolarized" U photons being changed superluminally into polarized photons of the V, H, D, or S variety, and hold out for a simpler explanation for the blue and green photon's identical behavior at distant calcites, an explanation that need not invoke eerie superluminal influences.

As an alternative model for the behavior of EPR-correlated photons, suppose an eccentric Texas millionaire mails a coin each day from Dallas to his son in London and to his daughter in Tokyo. He has such a large coin collection that his children cannot guess what sort of coin they will receive. But they do know that he always sends them identical coins in each mailing. Because of the millionaire's dependable habit, the instant his daughter opens her envelope in Japan and sees a gold coin, she becomes aware (faster than light?) that the coin her brother in England is looking at is also gold. Can the perfect EPR photon correlation be also explained the same way?

This correlated-coin model amounts to saying that the photons do not acquire their polarizations at the calcites but at the calcium-vapor source. The photons are not really without polarization after they leave the source, but like coins in an envelope, they each possess definite, but "hidden" polarizations—polarizations that reveal themselves to be identical when they strike their respective calcites. Although this sort of model is able to explain some of the EPR facts, it cannot explain them all. The gist of Bell's theorem is that all models of this type must fail to explain the EPR facts because such models do not involve real superluminal connections between individual quantum events. Bell showed that the blue and green photons from a calcium-vapor source are so strongly quantum-connected that only a model of quantum events that contains explicit FTL links—in this case, the instant long-distance transformation of a U photon into a V polarized photon—can ever hope to match the EPR facts.

The necessary FTL nature of quantum jumps in the EPR experiment is counterbalanced, however, by the equally necessary slower-than-light nature of quantum patterns. Each individual green photon changes its physical properties (from POP to PIP, for instance) with superluminal rapidity in agreement with Bell's theorem. However, in line with Eberhard's proof, no quantum patterns are ever observed to change faster than light. For example, the quantum pattern at Maxine's green calcite is a 50/50 random sequence of ups and downs no matter how Max sets his distant blue calcite. Although PIP photons are perfectly deterministic, equal numbers of H and V PIP photons yield equal numbers of up and down events. When the green photons are POP, on the other hand, their 50/50-random nature likewise produces an equal number of ups and downs at Maxine's green calcite. Whether Max is sending a "one" or a "zero," the pattern of quantum events at Maxine's green calcite is unchanged.

Since the quantum averages do not change when Max moves his calcite, Max's message is obviously not encoded in these averages. To discover Max's coded message, transmitted FTL in the altered character of each individual quantum jump, Maxine must muster the ability to measure the polarization of a single photon, and determine whether that photon is PIP or POP.

Sending messages backward in time would allow control of the present by manipulating events in the past. But the previous discussion shows that the secret of time travel via the quantum connection boils down to a single question: Can Maxine measure the polarization of a single photon? Never in the history of science has such immense power over physical events hinged on the ability to perform such a simple task.

The EPR experiment represents a giant step forward in FTL communications research. No need to worry about exotic schemes for producing tachyons, or for twisting spacetime into closed timelike loops. In order to send a signal faster than light today, all you, the reader, have to do is find a way to measure the polarization of a single photon. To make your job simpler, you are told beforehand that the photon is either *H, V, S,* or *D.* Moreover, to decode the green FTL message you do not actually have to measure the photon's polarization but merely decide whether it falls in the class (*H, V*) or the class (*S,D*).

Max sends Maxine an FTL message that she can decode using a single-photon polarimeter. But how does Maxine go about measuring the polarization of one photon? To get some idea of Maxine's plight, imagine a gambler who keeps ordinary coins in his right pocket and trick coins in his left—an equal mixture of two-headed and two-tailed silver dollars. He reaches into one pocket, pulls out a coin and flips it, catching it in the air and slapping it down on his wrist. "Heads," he calls. Maxine's dilemma resembles trying to guess which pocket the coin came from—and she must do this in a single observation. The gambler never flips the same coin twice; observing a single photon—by letting it trigger the up or down detector—destroys it. As a friend of mine once remarked: "Making quantum measurements is like testing flashbulbs—just when you find out they're good, they ain't." All quantum measurements have this one-of-a-kind quality. Each outcome is unique and can never be repeated.

The gambler's ordinary coins correspond to POP (random) photons encountering a calcite: they each have a 50/50 chance of going up or down. His trick coins correspond to PIP (determinate) photons: Each coin will always go up or always go down. However, provided the gambler only throws each coin once, the statistical behavior of fair coins from the right pocket is the same as that of trick coins from the left. Even though the physical situation is different in both the coin case and the photon case, the statistical outcome is always the same.

Suppose that Maxine discovers a way of cloning single photons, making copies that have the same polarization as the original, and performs independent measurements with her vertical calcite on each copy. (Using a photon cloning device amounts, in the gambling analogy, to persuading the gambler to throw the same coin more than once.) If the input photo was *H*-polarized, then all the copies would go up in a horizontal calcite. If the photon were *S* or *D*, on the other hand, only about half of the clones would go up. A photon cloning device would increase the amount of information one could get from a single photon and would allow an *H* or *V* photon to be distinguished from an *S* or *D* photon. The ability to make this distinction is all that you need in the EPR context to be able to signal faster than light.

A laser is an extremely intense source of light, which operates on the principle of "stimulated emission," an effect discovered by Einstein in 1917. When a photon enters a material full of excited atoms, such as an electrified gas, the photon's presence causes the atoms to want to emit light (stimulated emission) more intensely than they normally do (spontaneous emission). Unlike spontaneous light, which goes off in random directions with random polarization, stimulated light is emitted in the precise direction of the trigger photon and possesses exactly the same polarization. Stimulated emission is a kind of photon-cloning process. These cloned photons in turn clone further copies of themselves, causing an increasingly strong beam of identical photons to build up. In a laser, a pair of parallel mirrors reflects this amplified light back and forth through the excited gas. The repeated passage of identical photons through the gas takes further advantage of the medium's tendency to sprout identical photons in the presence of trigger light. This photon-cloning process is responsible for the high degree of spectral purity, the precise directionality, and the high intensity of the laser light source.

The heart of the FLASH (first laser-amplified superluminal hook-up) scheme is a laser gain tube arranged to clone Maxine's incoming green photon, making multiple copies which are then separately examined as to their polarization state. The FLASH scheme, proposed by the author in 1982, immediately focused physicists' attention on the question: Can a laser gain tube really duplicate photons in this way?

In 1983, Leonard Mandel at the University of Rochester worked out the theory of a simple gain-of-two laser amplifier at the one-photon limit and showed that although it did indeed clone photons, producing an identical copy of the input photons (a photon of polarization X produces two output photons each with polarization X), the gain tube also produces noise (in the form of spontaneous emission) that takes the form of a photon in the orthogonal polarization state $\bar{X}$.

According to Mandel's analysis, a perfect gain tube under the best of conditions will produce a faithful clone two times out of three. For a gain-of-two amplifier, faithful cloning means that when an H photon goes in, two H photons come out. However, one-third of the time the cloning process does not work and an orthogonally polarized photon comes out. In this case one *H* photon in gives one *H* photon and one *V* photon out. In other words, when operating at its theoretical limit, this photon copying machine malfunctions 33 percent of the time, but the rest of the time it correctly clones the input photon. If you had on hand one of these mostly reliable photon duplicators, could you measure the polarization of a single photon?

When the output of this dubious duplicator is analyzed with ideal calcites and detectors, it turns out that the Mandel gain tube offers no improvement at all for Maxine's photon discrimination problem. The effect of the noise photons exactly cancels the cloning advantage with the result that Maxine gets the same amount of information from the cloned photons that was available from a single photon. For instance, if the green photon goes up in her vertical calcite, she can say for sure that the photon was not *H*, but cannot make the crucial *H/V* versus *S/D* distinction. The FLASH scheme, trying to gain more information about a photon by duplicating it, simply doesn't work. This failure is not a peculiarity of the Mandel gain tube that some better photon amplifier might overcome, such limitations are an intrinsic feature of all quantum amplifiers.

The search for gravity waves is one of the frontiers of modern physics. These waves are so weak that amplifiers of the highest possible gain will be required to boost their energy to human-detectable levels. Motivated by the problem of how to amplify gravity waves maximally, CalTech physicist Carlton Caves examined the amplification process in general, and discovered that all quantum amplifiers produce a minimum noise that depends only on the amplifier's gain. The Mandel gain-of-two amplifier operates at Caves's limit, no better quantum amplifier is possible.

The search for FTL communicators in the cracks of present-day physics has been compared to the nineteenth-century search for perpetual-motion machines. In trying to understand clearly why perpetual motion machines invariably failed to work, physicists were led to the formulation of the first and second laws of thermodynamics which govern the amount and quality of energy available in any conceivable physical system. In a like manner, the study of why certain FTL schemes fail may also lead to certain general laws which on the surface seem to have nothing at all to do with the achievement of high-velocity communication, such as absolute gain limits on ideal amplifiers. Like the Kramers-Kronig relations discussed in Chapter 4, which result from imposing the speed-of-light limit inside transparent media, the impossibility of measuring the polarization of a single photon and the irreducible noise of ideal amplifiers are examples of new laws of nature that can be derived merely by assuming that the quantum connection cannot be exploited to send FTL messages.

All attempts so far to use the EPR experiment to send superluminal signals have failed. We know—courtesy of Bell's theorem—that an action taken by Max in Miami must have immediate physical consequences on Maxine's photons on Mizar. But Maxine, try as she will, just cannot seem to observe these consequences. What appears to be the case in the EPR situation is that Max can indeed send a superluminal message, but Maxine cannot manage to decode it, because Max's message is scrambled by perfect quantum randomness, by an unbreakable cipher to which only nature holds the key.

The attempt to send FTL messages via Bell's quantum connection can be compared to the attempt to utilize superluminal phase velocity for the same ends. In the quantum case, the message does not get through because to the receiver the signal looks utterly random. On the other hand, the superluminal phase wave is perfectly periodic—an infinitely long and infinitely boring sine wave, whose monotonously identical peaks and valleys are unblemished by any message. But perfect randomness and perfect order are equally devoid of meaning. Only a signal that lies somewhere between these two extremes qualifies as a meaningful message.

*   *   *

This EPR situation seems to show that nature undoubtedly uses superluminal links to accomplish her inscrutable ends but these deep quantum connections are private lines currently inaccessible—and perhaps permanently inaccessible —to humans for communication purposes. Bell's theorem shows that the world is built in a most curious fashion: To achieve merely subluminal effects, things are hooked together by an invisible underlying network of superluminal connections.

Although these invisible links connect point A to point B faster-than-light, they do not violate the causal ordering postulate. COP does not outlaw all FTL connections, but only those FTL connections that are "causal." The quantum connection escapes the COP prohibition because this connection is not "causal."

Whatever else it might be, a "causal" connection is at least asymmetric and controllable. It is asymmetric in the sense that one side of the connection is the cause, the initiator of the action, or sender of the message, and the other side of the connection is the effect, the receiver of the message or action. The quantum connection, as far as we can tell, is perfectly symmetric. Although I have dubbed Max the sender and Maxine the receiver of the FTL quantum connection, the results of the EPR experiment depend only on the relative positions of the two calcites. None of the calcites is special nor can be signaled out as sender or receiver. The outcome of the EPR experiment in all cases seems to depend upon the blue and green calcite setting in an entirely symmetric manner. Thus neither the blue nor the green calcite can be identified as the cause but each cooperates (apparently at FTL speeds) to bring about an overall outcome consistent with the quantum laws.

A "causal" connection must also be controllable. If I make a change at A, I should get a predictable change at B. If the change at B is unpredictable, how can I take credit for the change? Yet because of quantum randomness the changes that occur in the EPR experiment (whether a particular photon goes up or down) are totally beyond the control of the experimenter. The correlations are extraordinarily strong—no model in which connections obey the Einstein limit can explain why events at location A match so closely events at distant location B—but the individual events themselves are unpredictable.

Another example of an uncontrollable connection in physics occurs in John Cramer's advanced-wave model, discussed in Chapter 5. In Cramer's model, atom A (emitter) sends out a wave forward in time to atom B (absorber). When this wave arrives, atom B responds by sending a wave backward in time aimed straight at atom A. If this second process were independently controllable, we could routinely send messages backward in time. However in Cramer's model, the forward and backward wave are part of a single inseparable process—the two waves always occur in pairs and cannot be disentangled. Although the Cramer process is uncontrollable, it is certainly not symmetric —atom A (emitter) and atom B (absorber) play decidedly different roles in the exchange.

Because the quantum connection is neither asymmetric nor controllable, it does not qualify as a "causal" connection and is not forbidden by the causal ordering postulate. Thus even though the quantum connection is superluminal, it is not in conflict with special relativity.

Physicists call unmediated causal connections that jump from point A to point B instantaneously "action at a distance" and have generally rejected such unruly links as a basis for explaining what goes on in the world. No one has condemned action at a distance so strongly as Sir Isaac Newton, who wrote: "That one body may act upon another at a distance through a vacuum without the mediation of anything else . . . is to me so great an absurdity, that I believe no man, who has in philosophical matters a competent faculty for thinking, can ever fall into." Although many of Newton's ideas have been superseded, his distaste for action at a distance has remained an important guiding principle in physics.

Because it is not a "causal" connection, the quantum connection is not really an example of the generally despised action at a distance but represents a new and more subtle kind of long distance influence. Recently, Abner Shimony, professor of physics and philosophy at Boston University, proposed the term "passion at a distance" as an appropriate new name for the special type of superluminal link embodied in the quantum connection. In the EPR experiment, so it would seem, the blue and green photons are certainly united because of their previous association in the calcium atom, but not by anything so crude as "superluminal signals." Instead they find themselves entangled in a kind of "passion"—a passion uncontrollable, mutual, and so extraordinarily strong that it breaks through even Einstein's light barrier, the universal speed limit for more mundane relationships.

CHAPTER 9

Softening the Superluminal Paradox

"We're like millions of strands of spaghetti in the same pot. No time traveler can ever meet another time traveler in the past or future. Each of us must travel up or down his own strand alone."

"But we're meeting each other now."

"We're no longer time travelers, Henry. We've become the spaghetti sauce."

Alfred Bester
"The Men Who Murdered Mohammed," 1967

Superluminal speeds are certainly possible. Despite Einstein's famous prohibition, numerous physical processes routinely move faster than light. Likewise, certain equations that correctly predict the behavior of ordinary phenomena seem to imply the existence of faster-than-light or backward-in-time behavior as well. FTL loopholes undoubtedly exist, but what are the chances that we can exploit these loopholes to propel FTL starships or mediate superluminal communication networks? What the future will bring is anyone's guess. Here is my own estimate of which loopholes are live time tunnel possibilities and which are dead-end streets.

In a plasma or a waveguide the phase velocity of an electromagnetic wave exceeds the speed of light. However, since a phase wave does not carry energy, this superluminal action corresponds to no real physical movement. Furthermore, a phase wave is perfectly regular and utterly predictable, so it cannot carry a message. Phase waves have much in common with marquee lights which can also simulate a illusory sort of FTL motion. In my opinion the harnessing of phase waves for FTL communication is about as likely as using marquee lights for the same purpose.

Every wave, whether of water, sound, light, or quantum probability, obeys a certain wave equation. All wave equations possess two kinds of solutions: a "retarded" solution for which the wave's effect is felt after the wave is produced and an "advanced" solution where the wave's effect precedes the wave's production. Both retarded and advanced waves obey the Einstein speed limit—no FTL processes here—but advanced waves seem to offer the option (as would an FTL wave) of signaling backward into the past.

Physicists reconcile the existence of advanced solutions in their theories with the apparent absence of backward-in-time waves in the world in one of two ways. Either they simply throw away the advanced solution as a possibility that just does not happen in our Universe, or they assume (in the so-called "absorber models of radiation") that wave-mediated interactions really consist of equal parts advanced and retarded waves, but source and absorber properties cleverly conspire to cancel all advanced waves via destructive interference. To many physicists both tactics seem artificial, ad hoc attempts to force the equations to fit the phenomena. Here is a loophole some ingenious scientist might someday exploit, either by constructing an exotic system that produces free advanced waves or by utilizing "undulatory leftovers"—the uncanceled advanced waves of neutrinos or other weakly interacting radiation. The appearance of advanced-wave solutions in ordinary wave equations shows that the laws of physics do indeed permit backward-in-time wave motion, at least as a mathematical possibility. Perhaps it is only a human prejudice that the cause of wave motion must always lie in the past.

General relativity, Einstein's curved-space-based theory of gravity, enforces the Einstein speed limit locally but opens up the possibility of effective long-distance FTL travel via twisted geometry (space warps). Certain solutions of

Einstein's gravity equations (Gödel's universe, Tipler's cylinder, for example) describe space warps consisting of closed timelike loops (CTLs). CTLs can function both as travel routes to the past or as superluminal communication channels. CTLs are not as easy to ignore as, say, advanced-wave solutions. Physicists can easily discriminate against advanced waves because they are distinctly separate from retarded solutions. On the other hand, CTL solutions of Einstein's equations result from CTL-free solutions by a continuous change of parameters. Hence these CTL-infected space warps have the same existential status as the weakly warped space-time contours that stretch between Earth and the Moon.

CTL space warps may be mathematically possible but are they really achievable in practice? Studies so far suggest (but do not actually prove) that "you can't get there from here," that no manipulation of non-CTL space-time will ever result in the creation of a CTL. Tipler has shown that one cannot form a CTL without also producing a singularity somewhere, but no one has constructed an impossibility proof that CTLs can never be produced.

Computer simulations of rapidly spinning black holes have not yet produced any CTLs but because of the great complexity of Einstein's equations such models are necessarily oversimplified. In the real world there might well be unusual initial conditions that lead in a natural way to the formation of CTLs in space-time. Searching for ways to actually produce space warps seems a worthy use for fast computers. For decades curved space-time has been used for FTL travel in science fiction. As science fact the general relativity loophole remains in my opinion a live option for a future superluminal travel technology.

Perhaps computer models could uncover certain stable configurations of mutually orbiting Kerr black holes that create CTL pathways outside of their one-way membranes. More than a mere FTL signaling scheme, such a computer model would in effect present us with a roadmap for actual travel into the past. Although it might be many centuries before we could muster the advanced technology to actually place black holes in precision orbits, such a computer map would provide an irresistible motivation for the development of time-travel technology based on the manipulation of curved space in the vicinity of black holes.

Because it declares that every elementary event is fundamentally random, quantum theory liberates physics from the rigid determinism of Newtonian mechanics. The quantum revolution also opens up some FTL loopholes, permitting new kinds of particles to exist, particles that may travel faster than light (tachyons) or backward in time (antiparticles).

One of quantum theory's most puzzling features is that it describes the world when it is not observed in a decidedly different manner from the world when it is seen. Quantum theory represents all unobserved phenomena as probability waves—so-called "quantum wave functions"—but when phenomena are actually observed they manifest as particlelike events called "quantum jumps." A true quantum entity must be able to take on both a wave and a particle guise. Although physicists have constructed consistent theories of FTL particles, all plausible wave functions for such particles seem to obey the Einstein speed limit, thus the theoretical basis for tachyons is rather cloudy. Furthermore, not a shred of experimental evidence exists for these quantum FTL particles. In my opinion, the tachyon loophole is not an open path to a viable FTL technology.

When Paul Dirac wedded special relativity to quantum theory in the early thirties, his new relativistic quantum theory predicted the existence of an entirely new kind of matter—the antiparticles—whose existence has since been amply confirmed. Antiparticles are represented, in theory at least, as ordinary matter with negative energy that is traveling backward in time. Unlike tachyons, the existence of antiparticles is not in doubt. Can we then hitch a ride on an antiparticle and be carried backward into the past? Probably not. All phenomena so far observed seem to have positive energy. Despite its negative-energy representation, the positron (or antielectron) behaves like an ordinary positive-energy particle traveling in the ordinary forward-in-time direction. As far as its energy and time direction are concerned, antimatter looks to us just like ordinary matter. Positrons on the face of it seem to be no more useful than electrons as vehicles for travel into the past. In our positive-energy world everything travels forward in time; this ordinary behavior is just what Dirac's theory was designed to preserve. In my opinion, the backward-in-time nature of positrons is a mere theoretical artifact with no potential for time-travel applications.

The neutral kaon is a unique particle in the quantum zoo. Alone among elementary particles, the kaon undergoes reactions that violate the law of time reversal invariance. For example, a time-reversed movie of the reaction of $K_0 \to 2\pi$ would not look like a movie of the reaction $2\pi \to K_0$. The difference between these two movies is small—a few parts per thousand—but undeniable. There appears to be no way to exploit neutral K-decay for time travel but time travelers might be able to use this particle's special properties as a sort of temporal compass, to assess the direction of time's flow in unfamiliar universes.

In addition to the wealth of new particles and interactions that it permits, quantum theory ties its particles together in a particularly strong way. The mysterious "quantum connection" necessitated by Bell's theorem and demonstrated indirectly by the Clauser-Aspect experiment, links quantum systems that have once interacted but are now separate in a manner that is unmediated, unmitigated, and immediate. Eberhard's proof shows that this superluminal connection does not affect quantum averages but works instead on the individual event level, linking up single quantum jumps but not linking patterns of quantum jumps. Since, to a local observer, individual quantum events appear completely random, whatever message may have been superluminally sent from A to B is utterly undecipherable. The quantum connection seems to be a superluminal link accessible to nature alone—the world would work differently without it—but not as yet open to human control.

Since this superluminal link works only on the level of individual events, to be able to send superluminal messages via the quantum connection we must obviously understand more about "quantum jumps." The quantum connection, in my opinion, is a quite live FTL option because of physicists' general ignorance concerning the real nature of individual quantum events. The greatest unsolved problem in quantum theory is: What is a measurement? What is there about an "observation" that enables it to turn a wavelike possibility into a particlelike actuality? After more than a half a century of speculation we simply do not have a good answer to this fundamental question. Until we possess a clearer understanding of the quantum measurement problem, all theoretical prohibitions against using such measurements for FTL signaling rest on very shaky ground. Given our present state of ignorance concerning the nature of a quantum measurement, tapping into the network of FTL connections that links up distant quantum jumps all across the Universe seems to me to be a wide-open possibility.

As an index of my relative confidence in the FTL potential of these loopholes, if I were to invest $10 million in FTL research I would put an equal amount of money—perhaps $4 million each—into quantum connectedness and general relativity, plus $1 million for research into advanced potentials. This leaves $1 million to cover all the other options, not so likely candidates, in my estimation, for practical time-travel mechanisms.

Suppose this research pays off in a big way—not only can we send FTL messages but we can also physically travel backward in time. Because quantum connections and advanced waves are suited only for signaling, the ability to actually travel in time probably means that the space-warp research succeeded. One might imagine that when time-travel research succeeds once, it succeeds for all time; that if a time machine is invented in the future, it could be sent anywhere in the past including our own present. If this were true, one could argue that since time machines are not a common feature of our past and present, then no time machines will ever be invented. However certain obvious time-travel restrictions would invalidate this argument. If time travel is achieved within general relativity by constructing a region of space-time containing closed timelike loops (CTLs), intrepid temponauts can indeed travel into the past via a CTL, but they cannot go back to a time before the space warp existed. The date of such a machine's construction would mark a natural watershed in human history: all times B. T. M. (before time machine) being off-limits but all times A. T. M. (after time machine) open for exploration. Zero A. T. M. will mark the beginning of the Temponautic Era, in which time truly becomes another kind of space, all events equally accessible, whether in the past or in the future.

Because of the general consensus among scientists that time travel is impossible, few professionals have investigated the host of paradoxes that time machines might bring about. For the most wide-ranging studies of time-travel paradoxes, one must turn to science-fiction writers, who have imagined hundreds of time-travel puzzles and dozens of solutions to these paradoxes. These stories describe two kinds of time-travel paradox: the "consistent" and the "inconsistent" variety. In a consistent paradox, the time traveler goes back into the past to produce an event necessary for his own existence. In the inconsistent paradox, the time traveler triggers a past event that negates his present existence, such as killing his own grandfather.

In Mack Reynolds's "Compounded Interest," a man travels into the past and invests a few coins which, accumulating centuries of interest, yield the huge fortune he needs to build the time machine. No real paradox here, just a hint that money won't mean much to a time traveler. In some versions of the consistent paradox, the temponaut carries back plans for the time machine itself in order to insure its invention. But then who actually invented the time machine? Stories of this kind suggest that the origins of new ideas and inventions might lie partly in the future, that our intuitions of future events are shaped in part by the actual events themselves.

More problematic than ideas that come out of the blue are objects created out of nothing. In a typical story of this kind a geologist finds a futuristic artifact buried in an ancient geological stratum, places it in a museum, from which a time traveler steals it and deposits it millions of years in the past. In this sort of "consistent" time-travel story, the object is trapped in a closed temporal loop. It does not exist before its appearance in the rock; it ceases to exist after it is stolen from the museum. Since the time-looped object's origin is utterly mysterious, it is doubtful this kind of time travel can be truly called "consistent."

In the classic inconsistent time-travel paradox, a man travels back in time to kill his own grandfather, in which case he ceases to exist. But if he never existed, how can he kill his grandfather? The logical impossibility of such a situation is the strongest argument against time travel. This paradox must be faced squarely by any serious time-travel advocate and will surely be a cause for worry for the novice temponaut. As responsible advance agents for futuristic technology, science fiction writers have worked out several solutions to the kill-your-grandfather paradox.

In some stories, Gail Kimberly's "Minna in the Night Sky," for instance, the past can be observed by a temponaut but not changed. This solution eliminates paradox but the notion of an observation that does not change the thing observed violates the spirit of quantum theory. Quantum theory has abandoned the notion of an "immaculate perception," insisting instead that every observation, no matter how gentle, profoundly modifies the thing observed.

Many science-fiction writers employ "time police" to eliminate potential paradoxes "before they happen." In some stories the Universe itself behaves as the ultimate temporal traffic cop: in the vicinity of an incipient paradox, possible but highly unlikely events occur which abort the paradox without human intervention. Shortly after Frank Tipler published his *Physical Review* article describing how to build a time machine around a massive rotating cylinder, Larry Niven published a story in *Analog* magazine with the same title as Tipler's article: "Rotating Cylinders and the Possibility of Global Causality Violation." In that story, during an interstellar war, one side discovers some massive cylinders in space, artifacts of an ancient alien civilization. Realizing that time travel would give them an unbeatable military advantage, the generals decide to set the cylinders in motion. But the Universe, acting as time cop of last resort, has other plans up its sleeve.

The notion of waging war by going back in time is a common science-fiction plot device. In Fritz Leiber's "Change War" series, for instance, two eternally opposed factions, the "Snakes" and the "Spiders" struggle to modify the space-time landscape to their own advantage. In one of Leiber's stories, a change war veteran describes some of the misconceptions about space-time reconstruction: "Change one event in the past and you get a brand new future? Erase the conquests of Alexander by nudging a Neolithic pebble? Extirpate America by pulling up a shoot of Sumerian grain? Brother, that isn't the way it works at all! The space-time continuum's built of stubborn stuff, and change is anything but a chain reaction. Change the past and you start a wave of changes moving futurewards, but it damps out mighty fast. Haven't you ever heard of temporal reluctance, or of the Law of Conservation of Reality?"

The key to turning science fiction into science fact lies in knowing which aspects of relativity and quantum theory to take seriously and which to dismiss as mere empty furnishings. Dirac, for example, could have thrown away the negative-energy solutions to his relativistic version of quantum theory on the grounds that negative energies are plainly nonsensical since no particle can have an energy less than nothing. However by taking these solutions seriously—and reinterpreting them in a new way—Dirac was able to predict the existence of antiparticles before they were actually observed in the laboratory. The equations of physics are powerful sources of insight into new phenomena because they imply many more results than the few postulates and scant data on which they were originally based.

Special relativity, which in conjunction with the COP rule outlaws time travel, provides the main motivation for belief in time travel because relativity seems to assert that time is a dimension on a par with space. If time is like space then the past must literally still exist "back there" as surely as Moscow still exists even after I have left it. If the past still exists, then it makes sense to consider whether one could actually travel there.

The idea of time as a fourth dimension was a popular notion before Einstein was born. H. G. Wells used the fourth dimension as a plot device in his classic, *The Time Machine*, which was published several years before Einstein's paper on relativity or Minkowski's invention of space-time. As long as time as a fourth dimension remained only a popular notion, it was not a serious motivation for time travel. But as a crucial ingredient of a well-established physical theory, this notion now commands more respect. But should we take it seriously—and believe that the past really still exists—or dismiss the fourth dimension as a mere theoretical construct with no real physical implications?

If we take the fourth dimension seriously, we must believe that past and future have always existed, and that human consciousness, for reasons we do not comprehend, perceives this "block universe" one moment at a time, giving rise to the illusion of a continually changing present. In his *Philosophy of Mathematics and National Science*, mathematical physicist Herman Weyl described our limited perception of the four-dimensional world this way: "The objective world simply is; it does not happen. Only to the gaze of my consciousness, crawling upward along the life line of my body, does a section of this world come to life as a fleeting image which continuously changes in time."

If through some defect in human consciousness we do not perceive the real nature of time, we can imagine other beings not so constrained who experience the world as it really is. Kurt Vonnegut's fictional Tralfamadoreans in *Sirens of Titan* see the world this way. To a Tralfamadorean, human beings appear as "great millipedes with babies' legs at one end and old people's legs at the other." As yet no experiments inform us whether this aspect of relativity is literally true or not, but if we accept time truly as a fourth dimension, then time travel must be reckoned a serious possibility.

In the context of Newton's deterministic laws, special relativity described a block universe whose past and future were eternally fixed, a static museumlike world, whose entire history from beginning to end is changelessly frozen in amber. Because the foundations of quantum theory are not as well understood as those of Newtonian mechanics— nobody agrees on what a "quantum measurement" is, for instance—the quantum picture of the four-dimensional block universe is not so clear. Quantum theory suggests many different models of the four-dimensional landscape that relativity opens up to the daring temponaut.

We can get a feel for the strange way that quantum theory describes the world by comparing it to classical physics, which is made up of more commonsensical notions. The quantum world differs from the classical as Schrödinger's cat differs from Newton's cannon. Newton's mechanics described the flight of a cannonball in terms of a particular trajectory, determined solely by initial conditions and the laws of motion. Whether we look at it or not, the cannonball follows a single path through the air. Looking at the cannonball may change its motion a tiny bit but this change can be calculated and accounted for exactly. In Newtonian physics, looking at the cannonball does not radically change the cannonball's nature. In classical physics a "measurement" has no special status; a measurement is just an ordinary interaction.

Quantum physics, on the other hand, gives the measurement interaction a special status not shared by any other kind of interaction. Also, quantum theory describes the world in two entirely different ways, depending on whether it is being observed or not. When a system, be it a cannonball or a cat, is not observed, quantum theory describes it as a collection of possibilities, all of which move in a wavelike manner. Thus the quantum picture of an (unobserved) cannonball shows it traveling along many different trajectories at once, these different possibilities being symbolized by a mathematical object called the wave function. When the cannonball is looked at, however, it goes from many wavelike possibilities to one particlelike actuality—in a measurement-induced process called the "quantum jump." Thus a measurement, in the formalism at least, does not merely disturb the cannonball but profoundly changes its very nature, transforming it from a bundle of possibilities into a single actuality.

Quantum theory views the world as constructed out of two basically different kinds of interactions: ordinary interactions and measurements. Ordinary interactions, such as the action of gravity on the cannonball, may modify the wave function, changing and scrambling up its myriad coexistent possibilities, but an ordinary interaction can never turn a possibility into an actuality. It takes a measurement to do that.

In 1935, Austrian physicist Erwin Schrödinger dramatized the peculiar role of the measurement interaction in quantum theory by showing that this theory would, in principle, allow an unobserved cat to exist simultaneously in two possible states—as a live cat and as a dead cat—until some outside observer performed a measurement that promoted one of these two possibilities into an actuality. The peculiar status of Schrödinger's cat—dead and alive at the same time—has come to symbolize the bizarre nature of quantum theory.

In a nutshell, quantum theory describes the world as a collection of possibilities until a measurement makes one of these possibilities real. The import of quantum theory on time-travel theory depends on how seriously we take its picture of the unlooked-at world as mere possibility, and on what interpretation we place on these mysterious "quantum possibilities." One interpretation of quantum theory, devised by physicist Hugh Everett III for his doctoral dissertation at Princeton in 1957, is that each unobserved possibility represents a real universe. When a person makes a measurement, that person splits into as many different selves as there are measurement outcomes, and each of these selves dwells forever afterward in a separate universe. However, human consciousness works in such a way that it is only aware of one universe at a time. Just as the world is really four-dimensional but we don't perceive it that way, so also the world (according to Everett) really consists of a myriad of parallel universes all but one inaccessible to our merely one-track minds.

The Everett, or "many-worlds," interpretation of quantum theory, is most congenial to time travel. A time traveler can change the past but in so doing he is thrust into an alternative universe, saving the original timeline from paradox. Science fiction writers were using parallel universes to resolve time-travel paradoxes long before Everett's clever interpretation of quantum possibilities as actual worlds.

A measurement in the Everett interpretation does not really turn possibilities into actualities although it may look that way to a human observer. All "possibilities" are real to begin with. Instead, an Everett-style measurement entangles the multiple worlds of the observer—you, me, or Dr. Schrödinger—with the multiple worlds of the system—cat, cannonball, or quark—to bring about a myriad of new relationships in all of the Everett universes. It is a defect of our present form of consciousness that we remain oblivious of all these universes except the particular one we call "our own." As in so many other situations, full reality in the Everett picture is immensely larger than a simple human mind can perceive.

Many theoreticians are attracted to the Everett interpretation because it solves the quantum measurement problem by dethroning the measurement act. In the many-worlds picture a measurement is no different from the other interactions that make up the world(s). Most physicists, however, believe that the need to create billions of unobservable universes is too high a price to pay to solve the quantum measurement problem.

The majority of quantum physicists adhere to the Copenhagen interpretation, named after the hometown of Niels Bohr, the Danish physicist who invented and championed this view of what quantum theory really means. Bohr believed that despite the fact that the world deep down is undeniably quantum in nature, human beings can never directly experience the raw quantumness of the world. Our experiences must be forever "classical," that is, made up of clearly describable events that are distinct from one another. So far Bohr has been right. Even in the most sophisticated physics laboratory, the experiments themselves consist of quite commonplace events, clicks in counters, photographs of sparks and bubbles, or patterns of bits in a computer memory. The quantum world, on the other hand, is utterly indescribable in ordinary language. Bohr believed that it was a waste of time to even try to conceptualize this world since humans could never put such concepts to the test of experiment.

Bohr's colleague, Werner Heisenberg, was not so cautious. Heisenberg pictured the quantum world as consisting of "possibilities"—a mode of existence halfway between an idea and a fact. Many of these possibilities are contradictory—a cat dead, and a cat alive, for instance. These contradictions can peacefully coexist only because a quantum possibility does not fully exist but dwells in a shadowy limbo of semireality.

The bridge between the bizarre quantum world and the world of ordinary human life is a special kind of interaction called a "measurement." Only a "measurement" has the power to turn many shady quantum possibilities into one shiny real actuality. There is much controversy among quantum physicists over exactly what properties an interaction must possess before it is entitled to be called a "measurement." One influential wing of the Copenhagen school holds that the essence of measurement is "making a record," producing a permanent mark of some sort. A possibility is promoted to full reality status only when it generates a record somewhere, a trace accessible to public scrutiny. Some physicists maintain that when records are erased, the events that they had once recorded are returned to the status of possibilities. Because most events leave widely scattered traces, such erasures are extremely difficult to carry out. From this vantage point the quantum world looks like a vast sea of possibilities dotted with rare islands of fact, events that have managed somehow to achieve the archival quality that measurement status demands.

This record-oriented model of reality suggests that the past consists largely of possibilities, which could be realized by a time traveler as long as these new realities did not conflict with any existing recordings. Time travel in this view would be less like a single tape recording fixed for all time as in the Newtonian model, and more like a multitrack mix that could be overdubbed but never erased. History could be enriched by each time traveler "filling in the cracks" left by his predecessors. If erasures are disallowed as impractical (or dangerous), such a quantum-motivated model of time travel would seem to be paradox-free, changes in this kind of space-time regulated by a law of "conservation of reality" similar to Leiber's *Change War* rules.

This model of reality as a pattern of islands in a quantum sea, islands whose form and number are shaped in part from the future, recalls a speculation by I. J. Good concerning the supposedly random nature of quantum events. Most physicists believe that these quantum jumps are utterly random, that physics stops here and can go no further. Such events might appear random, Good proposed, if their causes were inaccessible to us because they originated partly in the future. Since Bell's theorem gives us reason to believe that these "random" events are FTL-linked perhaps these two barriers of contemporary physics are intertwined and will fall together: the twin beliefs that nothing can travel faster than light and that quantum jumps are inscrutably random.

Certain interpretations of quantum theory make the past appear more changeable than it seemed to be in the days of iron-clad Newtonian determinism. The quantum past, like the quantum future, may be open to revision via one of the many FTL or backward-in-time loopholes that uneasily co-exist in the equations of physics with the COP principle of special relativity that absolutely prohibits FTL signaling and transportation. Oddly enough, the COP rule does not outlaw backward-in-time messages—such as might be carried by advanced waves—as long as such messages in their backward flight do not exceed the velocity of light. Hence, as far as the present laws of physics go, there are really no theoretical barriers to sending messages into the past via advanced waves. The only drawback is experimental: No one has ever observed an advanced wave in nature nor has anyone any idea how to produce an advanced wave by clever manipulation of matter.

Likewise, the laws of general relativity obey Einstein's speed limit at every location, but permit effective faster-than-light travel via shortcuts in curved space. The quantum connection, which binds all quantum particles that have previously interacted by conventional means with an instantaneous but so far indecipherable communication link, may someday be harnessed in the service of human communication needs. No new physics is needed to argue that advanced waves, space warps, and the quantum connection are worthy candidates for the mainspring of a time machine. We may be sure that we have not discovered all the physics in the world; nature may contain more surprises in the area of time travel that conventional physicists have never dreamed of.

The rise of modern space science was largely inspired by

science-fiction stories about travel to other planets. We went there first in our minds, the rockets came later. Will time-travel stories and FTL physics possibilities ever converge to produce in some mind an idea for a practical time-travel device? Perhaps this compendium of FTL loopholes will hasten that process. It is my hope that someday we learn to evade the Einstein limit, to dissolve the old-fashioned distinction between past and future, and to make the whole space-time continuum our home.

RELATED READING

Technical Books and Articles

Bell, J. S. "On the Einstein Podolsky Rosen Paradox." *Physics* 1, 195 (1964).

Benford, G. A., Book, D. L., & Newcomb, W. A. "The Tachyonic Antitelephone." *Physical Review* 2D, 263 (1970).

Birch, P. "Is the Universe Rotating?" *Nature* 29, 451 (1982).

Bjorken, James D. & Drell, Sidney D. *Relativistic Quantum Mechanics*. New York: McGraw-Hill (1964).

Brillouin, L. & Sommerfeld, A. *Wave Propagation and Group Velocity*. New York: Academic Press (1960).

Carter, Brandon. "Global Structure of the Kerr Family of Gravitational Fields." *Physical Review* 174, 1559 (1969).

Carter, L. J., Ed. *Realities of Space Travel: Selected Papers of the British Interplanetary Society*. New York: McGraw-Hill (1957).

Caves, Carlton M. "Quantum Limits on Noise in Linear Amplifiers." *Physical Review* 26D, 1817 (1982).

Coleman, Sidney. "Acausality." In A. Zichichi (Ed.), *Subnuclear Phenomena, Part A*. New York: Academic Press (1970).

Clay, Roger W. & Crouch, Philip C. "Possible Observation of Tachyons Associated with Extensive Air Showers." *Nature* 248, 28 (1974).

Cramer, John G. "The Transactional Interpretation of Quantum Mechanics." *Reviews of Modern Physics* 58, 647 (1986).

Csonka, Paul L. "Advanced Effects in Particle Physics." *Physical Review* 180, 1266 (1969).

Datta, A., Home, D., & Raychaudhuri, A. "A Curious Gedanken Example of the Einstein-Podolsky-Rosen Paradox Using CP Nonconservation." *Physics Letters* 123A, 4 (1987).

Davies, P. C. W. *The Physics of Time Asymmetry*. Berkeley: University of California Press (1974).

Dewitt, B. S. & Graham, R. N. *The Many-worlds Interpretation of Quantum Mechanics.* Princeton: Princeton University Press (1971).

Eardley, Douglas M. "Death of White Holes in the Early Universe." *Physical Review Letters* 33, 442 (1974).

Eberhard, Philippe. "Bell's Theorem and the Different Concepts of Locality." *Nuovo Cimento* 46B, 392 (1978).

Feinberg, Gerald. "Possibility of Faster-than-light Particles." *Physical Review* 159, 1089 (1967).

Finkelstein, J. & Stapp, H. P. "CP Violation Does Not Make FTL Communication Possible." *Physics Letters 126A* 159 (1988).

Fitzgerald, Paul. "On Retrocausality." *Philosophia* 4, 513 (1974).

Froome, K. D. & Essen, L. *The Velocity of Light and Radio Waves.* New York: Academic Press (1969).

Ghirardi, G. C. & Weber, T. "On Some Recent Suggestions of Superluminal Communication Through the Collapse of the Wave Function." *Lettere al Nuovo Cimento* 26, 599 (1979).

Girard, Rejean & Marchildon, Louis. "Are Tachyon Causal Paradoxes Solved?" *Foundations of Physics* 14, 535 (1984).

Glauber, Roy J. "Amplifiers, Attenuators and Schrödinger's Cat." In Daniel M. Greenberger (Ed.), *New Techniques and Ideas in Quantum Measurement Theory. Annals of New York Academy of Sciences,* Vol. 480, p. 336 (1986).

Gödel, Kurt. "An Example of a New Type of Cosmological Solutions of Einstein's Field Equations of Gravitation." *Reviews of Modern Physics* 21, 417 (1949).

Harter, William G. "Galloping Waves and their Relativistic Properties." *American Journal of Physics* 53, 671 (1985).

Hawking, S. W., & Ellis, G. F. R. *The Large-Scale Structure of Spacetime.* Cambridge: Cambridge University Press (1973).

Herbert, Nick. "FLASH—A Superluminal Communicator Based Upon a New Type of Quantum Measurement." *Foundations of Physics* 12, 1171 (1982).

Kleinknecht, Konrad. "CP Violation and K_0 Decays." *Annual Reviews of Nuclear Science* 26, 1 (1976).

Mandel, L. "Is a Photon Amplifier Always Polarization Dependent?" *Nature* 304, 188 (1983).

Misner, C. W., Thorne, K. S. & Wheeler, J. A. *Gravitation.* San Francisco: Freeman (1973).

Morgan, J. W. "Superrelativistic Flight" *Spaceflight* 17, 252 (1973).

Morris, Michael & Thorne, Kip. "Wormholes in Spacetime and Their Use for Intersteller Travel: A Tool for Teaching General Relativity." *American Journal of Physics 56*, 395 (1988).

Ohanian, Hans. "Gravitation and Spacetime." New York: W. W. Norton (1976).

Partridge, R. B. "Absorber Theory of Radiation and the Future of the Universe." *Nature* 244, 263 (1973).

Pearson, T. J. & Unwin, S. C. et al. "Superluminal Expansion of Quasar 3C273." *Nature* 290, 365 (1981).

Penrose, Roger. "Naked Singularities." *Annals of the New York Academy of Sciences* 224, P125 (1973).

Pirani, F. A. E. "Noncausal Behavior of Classical Tachyons." *Physical Review* 1D, 3224 (1970).

Robinet, Loris. "Do Tachyons Travel More Slowly Than Light?" *Physical Review* 18D, 3610 (1978).

Sachs, R. G. "Can the Direction of Time's Flow Be Determined?" *Science* 140, 1284 (1963).

Schmidt, Helmut. "Can an Effect Precede Its Cause? A Model of a Noncausal World." *Foundations of Physics* 8, 463 (1978).

Shimony, Abner. "Controllable and Uncontrollable Non-locality." *Proceedings of an International Symposium on the Foundations of Quantum Mechanics, Tokyo* p. 225 (1983).

Stapp, H. P. "The Copenhagen Interpretation." *American Journal of Physics* 10, 1098 (1972).

Stapp, H. P. "Are Superluminal Connections Necessary?" *Nuovo Cimento* 40B, 191 (1977).

Taylor, E. F. & Wheeler, J. A. *Spacetime Physics*. San Francisco: Freeman (1963).

Tipler, Frank J. "Rotating Cylinders and the Possibility of Global Causality Violation." *Physical Review* 9D, 2203 (1974).

Tipler, Frank J. "Singularities and Causality Violation." *Annals of Physics* 108, 1 (1977).

Toll, John S. "Causality and the Dispersion Relations: Logical Foundations." *Physical Review* 104, 1760 (1956).

Wald, R. M. "Gedanken Experiment to Destroy a Black Hole." *Annals of Physics* 82, 548 (1974).

Wheeler, J. A. & Feynman, R. P. "Interaction with the Absorber as the Mechanism of Radiation." *Reviews of Modern Physics* 17, 157 (1945).

Wright, Philip B. "Immediate Interstellar Communication." *Speculations in Science and Technology* 2, 211 (1979).

Wooters, W. K. & Zurek, W. H. "A Single Quantum Cannot Be Cloned." *Nature* 299, 802 (1982).

Zajonc, Arthur G. "Proposed Quantum Beats, Quantum-Eraser Experiment." *Physics Letters* 96A, 61 (1983).

Popularizations

Bilaniuk, Olexa-Myron, & Sudarshan E. C. George. "Particles Beyond the Light Barrier." *Physics Today* p. 43 (May 1969).

Compton, Tony. "Our Tomorrows Are All Their Yesterdays." *New Scientist* p. 700 (June 26, 1975).

Crease, Robert P. & Mann, Charles C. "The Physicist that Physics Forgot: Baron Stückelberg's Brilliantly Obscure Career." *The Sciences* p. 18 (July/August 1985).

Dettling, J. Ray. "Time Travel: The Ultimate Trip." *Science Digest* p. 81, (Sept. 1982).

Donaldson, Thomas. "How to Go Faster Than Light." *Analog Science Fiction/Science Fact* p. 77 (June 1985).

Gardner, Martin. "Can Time Go Backward?" *Scientific American* p. 98, (Jan. 1967).

Good, I. J. "Two-Way Determinism." In I. J. Good (Ed.) *The Scientist Speculates* New York: Capricorn p. 314 (1965).

Gribbin, John. *Spacewarps.* New York: Delacorte (1983).

Grigonis, Richard. "Sixth Generation Computers." *Dr. Dobbs Journal* p. 37 (May 1984).

Guinness Book of World Records. New York: Bantam (1986).

Herbert, Nick. *Quantum Reality: Beyond the New Physics.* New York: Doubleday (1985).

Kaufmann, William J., III. *The Cosmic Frontiers of General Relativity* Boston: Little, Brown (1977).

Livingston, Dorothy Michelson, *The Master of Light: A Biography of Albert A. Michelson.* New York: Charles Scribner's Sons (1973).

Muller, Richard A. "The Cosmic Background Radiation and the New Aether Drift." *Scientific American* p. 64 (May 1978).

Nicholson, Ian. *The Road to the Stars.* New York: William Morrow (1978).

Niven, Larry. "The Theory and Practice of Time Travel." *In All the Myriad Ways.* New York: Ballantine (1971).

Rothman, Milton. "Things That Go Faster Than Light." *Scientific American* p. 142 (July 1960).

Rucker, Rudy. *The 4th Dimension: Toward a Geometry of Higher Reality*. Boston: Houghton Mifflin (1984).

Science Fiction

Anderson, Poul. *Tau Zero*. New York: Berkley (1970).

Aldiss, Brian. *Starswarm*. New York: New American Library (1964).

Asimov, Isaac. "The Endochronic Properties of Resublimated Thiotimoline." In *The Early Asimov*. New York: Doubleday (1972).

Ballard, J. G. "Time of Passage." In J. D. Olander & M. H. Greenberg (Eds.), *Time of Passage*. New York: Taplinger (1978).

Benford, Gregory. *Timescape*. New York: Pocket Books (1980).

Bester, Alfred. "The Men Who Murdered Mohammed." In Robert Silverberg (Ed.), *Voyagers in Time: Twelve Stories of Science Fiction*. New York: Hawthorn Books (1967).

Dann, Jack & Zebrowski, George (Eds.). *Faster Than Light*. New York: Ace Books (1976).

Effinger, George Alec. *The Bird of Time*. New York: Doubleday (1986).

Gerrold, David. *The Man Who Folded Himself: The Last Word in Time Machine Novels*. New York: Random House (1973).

Kimberly, Gail. "Minna in the Night Sky." In Roger Elwood (Ed.), *The Far Side of Time*. New York: Dodd, Mead (1974).

Le Guin, Ursula K. *The Dispossessed*. New York: Avon (1974).

Le Guin, Ursula K. *Rocannon's Way*. New York: Harper & Row (1977).

Leiber, Fritz. *The Big Time*. New York: Ace Books (1961).

Leiber, Fritz. "Try and Change the Past." In Robert Silverberg (Ed.), *Trips in Time*. New York: Thomas Nelson (1977).

Nelson, Ray. "Time Travel for Pedestrians." In Harlan Ellison (Ed.) *Again, Dangerous Visions*. New York: New American Library (1972).

Niven, Larry. "Rotating Cylinders and the Possibility of Global Causality Violation." In *Convergent Series*. New York: Ballantine (1979).

Robinson, Spider. "Half an Oaf." In *Antinomy*. New York: Dell (1980).

Rucker, Rudy. "Jumping Jack Flash." In *The 57th Franz Kafka*. New York: Ace SF (1983).

Sheckley, Robert. "Specialist." In C.W. Sullivan (Ed.) *As Tomorrow Becomes Today*. New York: Prentice-Hall (1974).

Vonnegut, Kurt, Jr. *Sirens of Titan*. New York: Dell (1970).

Watson, Ian. *The Very Slow Time Machine*. New York: Ace Books (1979).

Wells, H. G. *The Time Machine*. New York: Bantam (1982). (In print for almost a hundred years. Originally published in 1895.)

Wilson, Robert Anton. *Schrödinger's Cat*. New York: Pocket Books (1979).

Zelazny, Roger. "Divine Madness." In Robert Silverberg (Ed.), *Trips in Time*. New York: Thomas Nelson (1977).

INDEX